HEM
BETHL
HEM
BETH
HEM
BETHL
HEM
BETH

A SKYSCRAPER GOES UP

A fascinating account of building a 54-story office tower—from the first plans to the finishing touches.

STANDARD OIL COMPANY

A SKYSCRAPER GOES UP

by Carter Harman

illustrated with photographs

Random House New York

For
Lisa, Bruce, Scott and Alex

 Published in the United States by Random House, Inc., New York, and simultaneously in Canada by Random House of Canada Limited, Toronto. Manufactured in the United States of America *Designed by Murray M. Herman*

Library of Congress Cataloging in Publication Data
Harman, Carter. A skyscraper goes up. SUMMARY: A step-by-step account of building a fifty-four story office tower from planning to a view of the completed structure. 1. Skyscrapers—Juvenile literature. [1. Building. 2. Skyscrapers] I. Title. TH1615.H37 690.5′23 72-11059. ISBN 0-394-82147-5 ISBN 0-394-92147-X (lib. bdg.)

I wish to thank the specialists from the following companies, who gave generously of their time and information: Exxon Corporation; Rockefeller Center, Inc.; Harrison & Abramovitz & Harris, architects; Syska and Hennessy, Inc., mechanical engineers; Edwards and Hjorth, structural engineers; the Bethlehem Steel Corporation; the George A. Fuller Company, general contractor; and Civetta Excavating, Inc.

CARTER HARMAN

CONTENTS

A SKYSCRAPER GOES UP

An excavation in the middle of the city, so large that the trucks inside it seem to be toys.

A SKYSCRAPER GOES UP

It was a chilly December morning in 1968. Office workers were scurrying along the windy sidewalks of midtown Manhattan, or dodging the taxis and buses that jammed the streets between towering skyscrapers.

A visitor stepped off the sidewalk, through a wide gate in a temporary wooden wall—and into another world. Instead of looking up at a tall office building, he was looking down into a vast hole. It was the excavation that would soon contain the newest skyscraper in Rockefeller Center, planned as the corporate home of the Exxon Corporation, formerly known as the Standard Oil Company (New Jersey).

Picking his way down the muddy ramp, he realized that he should have been wearing protective earplugs. The "normal" roar of New York City was completely drowned out by the clamor that bounced back and forth between the rocky walls of the hole.

There were machines pumping, digging, drilling, clawing, and carting. They made roaring, hissing, grinding, crunching, and ripping noises. The thundering engines of sixteen air compressors, two big diesel cranes, three power shovels, and assorted crawling bulldozers, back-hoe diggers, drills, and front-end loaders blended into one huge roar. Even above their din, jackhammers clattered like giant machine guns.

Near one wall of the excavation, the shrill toot of a whistle was heard, followed by two more toots. All the workers moved away from the area. Then came a dull *whump!* The ground trembled for a moment and a cloud of dust rose. A blast had been completed.

The visitor found the noise so overpowering that he could not even hear the roar of a blunt red-and-green dump truck coming right at him up the ramp. It was fully loaded with ten tons of broken rock and earth, and its rear end slid back and forth in the mud as it climbed the steep grade.

The visitor could easily guess what the machines were doing. Their job was to dig a hole 450 feet long, 200 feet wide, and about five stories deep, most of it out of solid rock. They had to square off the sides of the excavation so that smooth concrete walls could be built there, pump out any water seeping into the hole, and brace the walls so that the sidewalks and buildings along the edge of the site would not cave in.

In addition to the machine operators, there were many other workers, wearing colored plastic helmets. Some carried rolls of blueprints, some sighted through surveyors' telescopes, and others did jobs that were not immediately understandable. It was hard to believe that out of such chaos a huge skyscraper would be created. The visitor wondered how it would be done.

More than three years later, the same visitor found his way up to the roof of the completed Exxon Building. High above the roar of the city, all was quiet. The sun shone brightly and a brisk wind blew from the west. Far below, the streets seemed as narrow as footpaths and the taxis and buses looked like crawling insects. Pedestrians were hard to see at all.

Looking out toward the horizon, the visitor found it hard to believe he was standing 750 feet above the street. The new building was tall, but many others were even taller. For this

While men at the bottom left prepare to blast, power shovels and pneumatic drills (background) keep up the din.

Three years later: the Exxon Building.

was New York City, which has the largest collection of skyscrapers in the world.

Far to the south, he could make out the outlines of the tower of the Woolworth Building. It was nearly sixty years old, and its sides were "set back" at intervals to give it a graceful, tapered effect. Yet it reached a height of 792 feet, a few stories higher than the Exxon Building itself.

Closer at hand, to the southeast, were the pointed spires of the Chrysler and Empire State buildings. Both of them were more than forty years old, but the visitor could easily see that they were taller than the building he stood on. The Chrysler Building towered over 1,000 feet, and the long, slender television antenna atop the Empire State stretched up more than 1,200 feet. Still nearer stood the six-sided Pan Am Building. This giant structure had a flat top which had once been used as a landing spot for helicopters.

Even the building directly across the street to the east, the RCA Building, rose higher than the Exxon. It was nearly forty years old and had 70 floors. From his vantage point, the visitor could see office workers at their desks, more than 700 feet above the ground.

All of these huge towers provided working space for thousands of businessmen, executives, clerks, and secretaries. Some of the skyscrapers housed more people than live in a small town. And they provided more than just space—heat in winter, cool air in summer, light, running water, and everything else necessary for the workers' comfort and efficiency.

Not all of the buildings were old. Far to the south, near the tip of Manhattan Island, the visitor could see the twin towers of the World Trade Center, each 110 floors high. They were just being completed, and tenants were already moving in. Immediately across the street to the south, two other buildings were still under construction. They would have 45 and 51 floors and would be nearly as tall as the Exxon Building.

Looking south from the RCA Building: new skyscrapers rise in the foreground as the Empire State Building dominates the scene. In the distance, nearly four miles away, are the twin towers of the new World Trade Center.

New York has more tall buildings than other cities, but skyscrapers are going up in cities all over the world. As cities become crowded, big companies begin to build skyscrapers to house their office workers near the center of town. No matter where they are and no matter how different they look, all of these giant structures are built by similar principles and techniques.

The visitor took one more look around from the top of the Exxon Building, then headed back inside for a quick ride down to ground level. He had a special feeling for this building, for he had been following its construction from the start. Watching the structure grow, he had learned how a skyscraper is built.

PLANNING The Exxon Corporation is one of the world's great corporations. It has offices in major cities around the world, and in the United States its products—gasoline, fuel oil, and many others—are bought by millions of customers.

Exxon employs many thousands of workers. Among them are geologists to look for oil; drillers and pumpers to get the crude oil out of the earth; sailors to carry it across the sea in tankers; scientists and engineers to refine and convert the oil into usable products; and salesmen to sell these products. The company has a headquarters staff of more than 2,000 people just to direct and keep track of all its operations.

Exxon has had its headquarters in New York City from its beginning. Its first offices were "downtown" near the southern tip of Manhattan Island. In the 1940s it moved "uptown" to Rockefeller Center, a cluster of large office buildings in a newer part of the city. One of the newer Rockefeller Center buildings was named the Esso Building, after one of Exxon's earlier trade names.

But the company continued to grow. Soon its staff was so large that it occupied offices in several Rockefeller Center buildings. During the 1960s it became clear that the company ought to have new quarters where most of its staff could be under one roof.

The first question was where to locate the new headquarters building. Many major companies have decided that the city—any city—is just too crowded. They have built large, rambling buildings in parklike locations many miles from the city. But there are many advantages to having offices in the city, and Exxon finally decided to remain in New York and to stay in the same neighborhood.

This decision meant that the company would have to build a skyscraper. Land in the center of large cities is so scarce and so expensive that even a large corporation cannot afford to buy a huge lot and build a rambling building only a few stories high. It would be much less expensive to use a smaller plot of land and build straight up.

But building a skyscraper is a specialized business, and Exxon decided to go into a joint undertaking with Rockefeller Center, Inc., the company that had built and managed the buildings in Rockefeller Center. Exxon and Rockefeller Center would plan the building together and share the cost of construction. Rockefeller Center would manage it and Exxon would use much of the building for its headquarters staff. The remaining space would be rented to other businesses and the two partners would share the profits.

The original buildings of Rockefeller Center were built in the 1930s and 1940s. Later, Rockefeller Center, Inc., gained control of more land just across the street from the original buildings. In the late 1950s the Center, in partnership with Time, Inc., built the Time & Life Building on part of the new land. And in the late 1960s it was decided to use another piece of the land, right next to the Time & Life Building, for the new Exxon headquarters.

Exxon officials felt that this piece of land would be a fine location for the new skyscraper. It was one of the most desirable business locations in the city. While conducting the planning for the Exxon Building, Rockefeller Center, Inc., also made agreements with other corporations to erect two more skyscrapers nearby.

The planning for the Exxon Building was a complicated job, involving economics, architecture, engineering, law, and many other specialties. One of the first questions that had to be answered involved money. How big a building could the companies afford to build? Of course, there had to be enough room for all the offices needed by Exxon. Additional offices could be rented out, and the rent would help pay for the cost of the building. But if the building were too large, it would not only be very expensive to build, it would also be difficult to fill and manage. The two companies finally decided how big the building should be and how much it should cost.

Then they sat down to discuss their plans with an architect. They had chosen the firm of Harrison & Abramovitz & Harris. Mr. Wallace K. Harrison, one of the firm's partners, had helped design the original Rockefeller Center buildings forty years earlier. As soon as the architects knew what the owners needed, they could begin to envision the building. Then the next step would be to prepare drawings, plans, and specifications for the new structure. The owners might ask for changes, but eventually the architect and owners would agree on how the building should look, how much floor space it should have, and all the other details.

One thing the architects had to consider when drawing up their plans was the laws that control new buildings. New York, like other cities, has divided its land into *zones.* A builder may not go to a quiet neighborhood and build an office building, because land there is zoned for residential use only—for homes or apartments. The site for the Exxon Building is zoned for commercial use—it may be used for an office building. The land is also approved for high-rise buildings, which means that a skyscraper may be built there.

The city's Building Code provides many special rules for

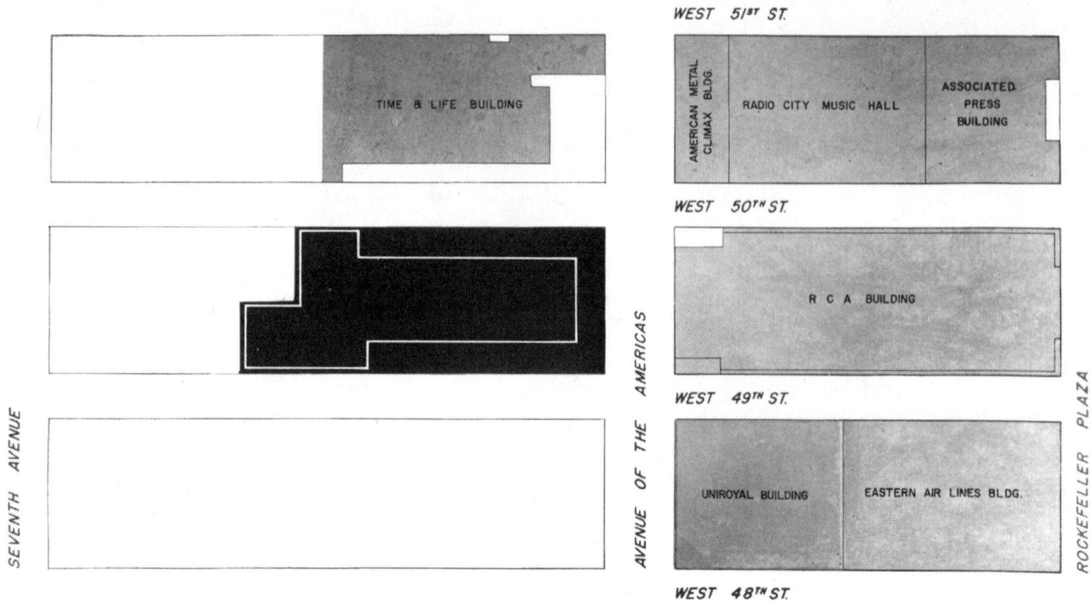

The black rectangle shows the site of the new Exxon Building, just south of the Time & Life Building and just west of the RCA Building.

skyscrapers. These rules are designed to protect those who will work in the building and others who work nearby. They specify how strong the building has to be and what equipment it must have. In addition, the Code specifies that a building of this height be set back a certain distance from the street. This is to make sure that people on the ground level in the canyons between tall buildings can still enjoy some open space, sunlight, and fresh air.

Architects also must consider the appearance of a new building. Hopefully, when it is finished it will be attractive as well as useful. Harrison & Abramovitz & Harris had a particular problem with the Exxon Building. They felt that a new structure should look modern and "light." But since the building would be part of Rockefeller Center, they felt it should harmonize with the older, massively styled buildings across the street.

The architects' finished drawings showed a spectacularly

An old aerial view of Rockefeller Center has been retouched to "make room" for the Exxon Building, then in the planning stages. The RCA Building stands out at center and the Time & Life Building is partly hidden by the phantom of the new structure.

beautiful building that would stand 750 feet tall and have 54 floors. The outside walls featured a series of dramatic vertical limestone columns. (These would match the limestone used in the original Rockefeller Center buildings.) Between each column was a shiny strip of glass that seemed to stretch from the ground to the very top of the building. The cost of such a building—for the services of architects and engineers, for materials and labor—would run many millions of dollars, perhaps as much as a million dollars per floor.

When their renderings and general specifications were approved, the architects began the detailed planning. They were assisted by other planners who specialize in engineering.

The structural engineers helped decide how the building the architects envisioned could be made safe and durable. They estimated its total weight and considered other pressures on the building—wind and the weight of people and machinery. (If the building had been located in an earthquake region, they would have had to consider the possible effects of severe tremors.) They also calculated how strong each major element of the building had to be in order to withstand these pressures.

Then the architects and engineers completed the detailed drawings that would guide all the workers when construction actually began. These drawings were reproduced as *blueprints,* large maplike sheets on which all the details of the building have been carefully worked out.

There is no single set of blueprints for the whole building. Instead, separate sets are made for each subcontractor. One set is for the excavators, one for the steelworkers (or "ironworkers," as they are called, after the original name of their union), one for the concrete workers, one for the electricians, one for the plumbers, one for the heating and ventilating contractor, one for the elevator workers, and so on. The ironworkers' plans show where every column, brace, beam, truss, and other steel member is to fit, and even how each is to be fastened to the others. The electricians' plans show where all the light sockets, lights, and switches will be and how they are to be connected. The plumbers' plans show where the pipes, toilets, and basins are to be located. The air-conditioning contractor's plans show where all the ducts for warm and cool, fresh and used air are to be hidden behind walls and ceilings.

Once these plans have been completed, the builders engage a *general contractor,* who will be responsible for construc-

An architect's model shows how the front of the Exxon Building will look when complete. A pool separates the building from the Avenue of the Americas.

tion according to the plans. The general contractor sends detailed plans to *subcontractors,* and they are asked to *bid* on the part of the job they are prepared to do. For example, several different plumbing companies may bid for the plumbing work in the building. The company that offers to do the work for the smallest amount of money usually gets the job.

The subcontracting companies then begin to order the materials that will go into the building—millions of dollars' worth of steel, cement, gravel, stone, pipe, wire, fireproofing

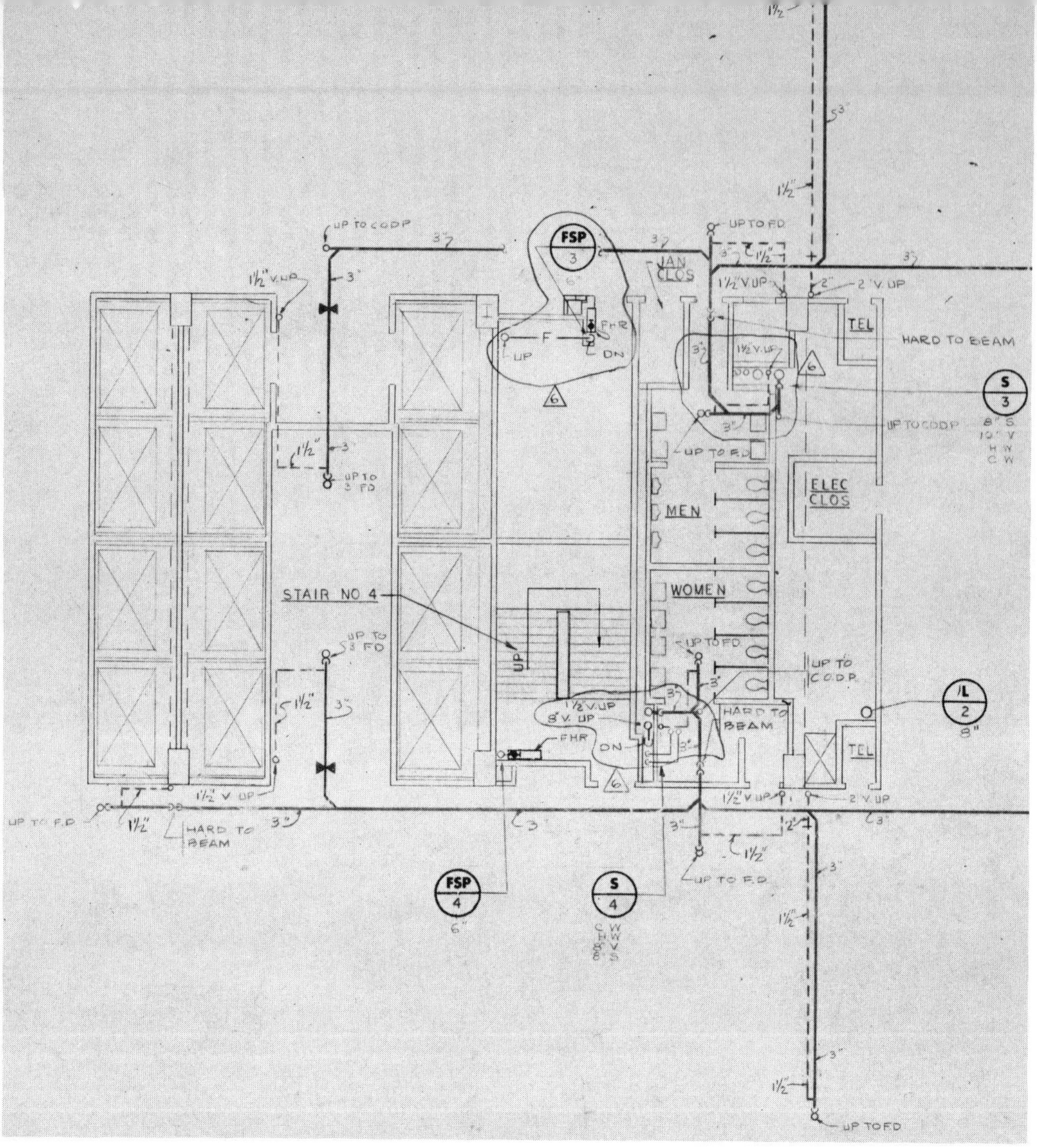

A small part of one of hundreds of blueprints for the building. Elevator shafts are marked by large X's on the left. A stairway can be seen at center, and toilets are at right center. The circles and arrows identify minor revisions in the plans.

materials, and so on. Again, different manufacturers may bid to supply the materials, and the lowest bidder generally makes the sale.

While the contractors were being lined up and the materials ordered, preparations were being made to clear the land and get it ready for construction.

DEMOLISHING AND DIGGING

The site on which the Exxon Building would stand was a rectangle bordered on three sides by city streets. The front of the building would face east onto the Avenue of the Americas, taking up the whole block between 49th and 50th streets. The long sides of the structure would stretch more than 300 feet along 49th and 50th streets. This land was not a vacant lot. In midtown New York, as in the heart of other large cities, there is no vacant land. So if a new building is to go up, old buildings must first be torn down.

Rockefeller Center, Inc., already owned the land and the buildings. Now it was necessary to deal with the people there. The old buildings contained both shops and apartments, and all the tenants had to move out. In some cases, shopkeepers and residents could move to new and better locations. But some of them didn't know where they could go. In such a situation special relocation companies are often used to make sure that tenants move out when their leases expire and to give them help in finding other places to live or set up shop.

One of the first contractors to go to work was a demolition company. As soon as one of the buildings became vacant, men were sent in to tear it down. Some of the buildings were 100 years old and were in bad condition. But old buildings often contain many valuable parts. There may be marble sink

The site of the new building before the old one- and two-story structures were demolished. The Time & Life Building is at upper right.

tops and fireplaces, and antique moldings and door handles. Demolition men take these parts first and are careful not to damage them, since they can be resold.

Early in the demolition work, the men removed dozens of heavy old doors from the buildings and placed them in a row alongside a scaffolding above the sidewalk to protect passers-by from falling debris. Then they went to work with sledge hammers and crowbars to break up the roofs and floors and knock down old brick walls.

Even before the buildings had been leveled, surveyors went to work checking the dimensions and boundaries of the site. Boundary lines may look perfectly clear on a map, but sometimes—especially in old neighborhoods—the corner of a building may actually rest on another owner's lot. In order to

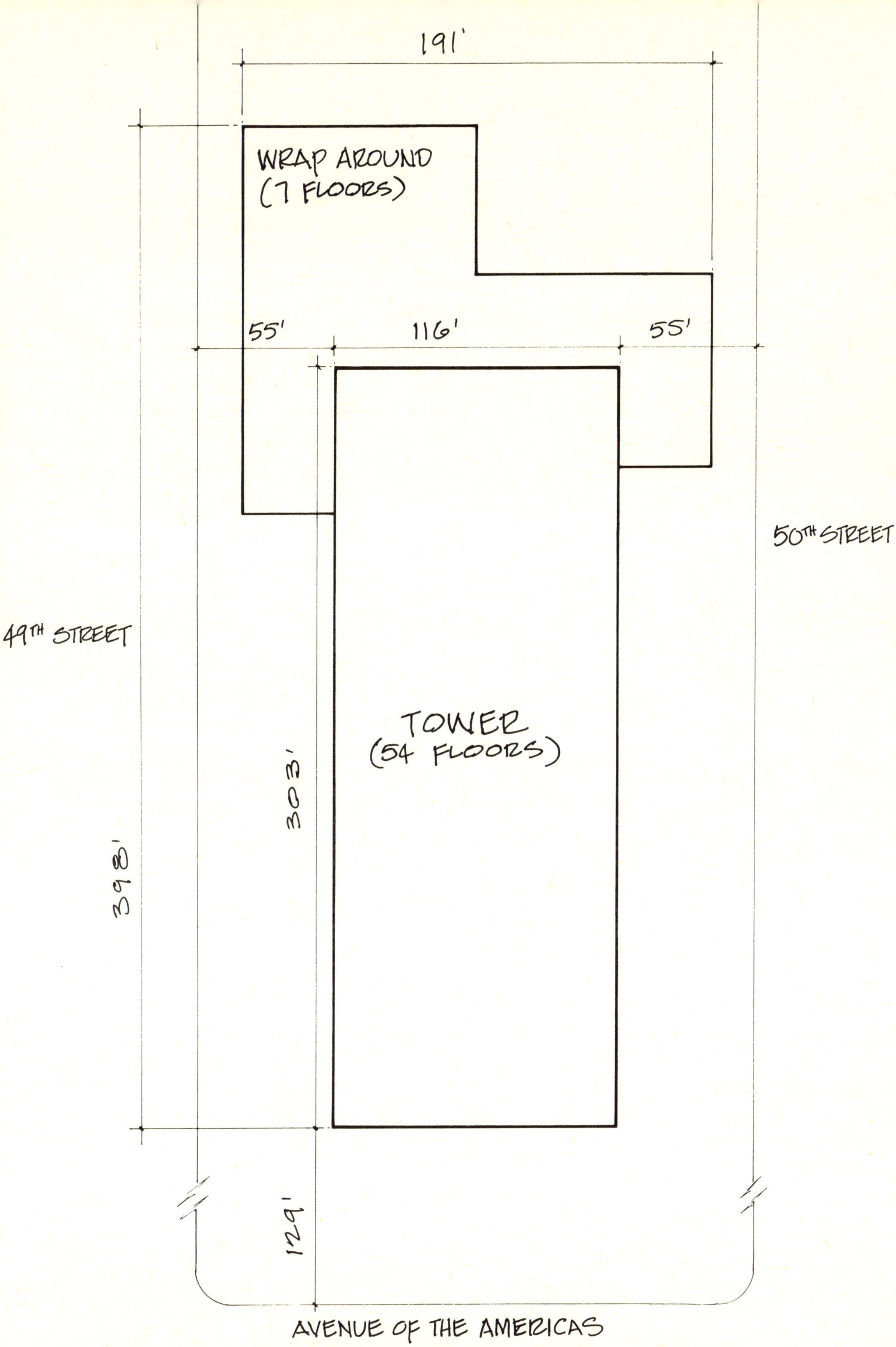

This diagram shows the dimensions of the new skyscraper: nearly 400 feet long and 200 feet wide at its longest and widest points. The tower section, which will be over 700 feet high, is more than 300 feet long.

avoid disputes with neighbors, a builder must know the exact boundaries of the land.

The owners also confirmed their legal title to the site. If it were discovered after the huge skyscraper was built that some small piece of the land was owned or controlled by someone else, this person might go so far as to demand that the building be torn down. More likely, he would sue the building's owners. To avoid this possibility, special companies study the ownership of building sites in order to *guarantee* the builders' title to it.

The next step was to make an underground study of the site. Old city and utility company maps showed the location of water, gas, and electric connections which would have to be sealed off. In addition, investigators on the site looked for old cisterns and forgotten tunnels that might cause dangerous cave-ins.

Another important task was to sample the ground itself. Before preparing the foundation of the new building, the excavators needed to know what kinds of soil and rock lay under the site. A crew of specialists penetrated the ground with a hollow drill, which extracted solid, cylinder-shaped samples of earth known as cores. These cores provided an exact picture of the different levels of earth. By drilling in a pattern, the specialists could prepare a map of the underground, showing clay, loose rock, and finally the solid *bedrock* that would support the building.

In New York City, the bedrock is an extremely hard rock called Manhattan schist. As it happened, the schist on most of the Exxon site lay very near the surface. Instead of digging through many feet of clay to find firm support for the huge building, the excavation contractors would have to dig out a hole in the solid rock itself.

This meant there would be blasting—and lots of it—to crack and shatter the rock into pieces small enough to be

Excavation is beginning. The building at the rear houses a Chinese restaurant, the "China Fair." The workmen are preparing holes for blasting.

trucked away. Once again, the way had to be prepared. Any blast powerful enough to split Manhattan schist would be powerful enough to shake the neighboring buildings, and possibly crack their walls or rupture their ancient water and gas pipes.

To prevent such damage, the excavating company sent still another team of experts into every neighboring basement to study and record its condition before blasting. Guided by their findings and by city laws, the experts limited the amount of explosive that could be used in various parts of the site. As a further precaution, a honeycomb of holes could be drilled in the ground near the base of neighboring brick buildings to help cushion the shocks.

When the preparations had been completed, the site was a level stretch of strewn bricks and rubble. The excavating contractors had already surveyed the site, and they knew the exact size and location of the hole. Now they were ready to dig.

On August 5, 1968, the excavation was begun. Gradually the trucks, mechanical shovels, jackhammers, and other equip-

Some months later, the "China Fair" towers over the excavation, now some 60 feet deep. The huge brace running diagonally from the ground to the wall helps keep the building from falling into the hole. Workmen on the scaffolding are preparing to make a concrete wall like the one at the right.

ment arrived to help with the noisy work of digging into solid rock. As we have seen, the growing excavation was soon a beehive of activity. But the most important job was done by the blasting teams.

For the blasting they used an explosive related to dynamite. If a stick of this explosive were set off in the open, the shock wave might knock bystanders off their feet and break windows, but the blast would scarcely scar a solid rock. To damage rock, each stick must be buried deep in a hole in the rock and held there by some kind of wadding. That way, the force of the explosion causes the rock to crack and open up.

Before each blast, several automatic drilling machines must drill holes in the rocks. The machines have long, pointed bits that extend downward from a skeletal framework on the front of a four-wheel carriage. They are wheeled into position and then the bits drill slowly into the rock. The location of each hole and its angle through the rock are carefully planned to get the best results from each blast.

The explosive itself is not very sensitive to normal jolting, so it must be set off by a severe shock. This is provided by a small blasting "cap" which is filled with another type of explosive that can be set off by an electric spark. When a set of holes is ready, the explosive-handler inserts a pointed blasting cap into each stick of explosive. He carefully loops the electrical wire from the cap around the stick so that it will not get pulled away. Then he slides the stick into the hole, leaving only the wire visible. Another man pushes the explosive down to the bottom of the hole with a pole that looks like a very long broom handle. Then the rest of the hole is filled with wadding—long, tube-shaped bags filled with fine sand.

Several holes were prepared in this way. From each of them came a wire that was attached to a small box far from the blast. This box would provide the electrical impulse to set off the several sticks of explosive all at once. Then a diesel shovel rumbled in, swinging five or six large mats woven out of heavy steel cables. It flopped these down onto the blast area to contain the flying debris.

A long toot on what sounded like a tugboat whistle warned that the blast was about to take place. Everybody moved back to a safe distance. Two more toots gave final warning. After a pause came that dull *whump!* It shook the ground and could be felt by office workers for blocks around. The heavy mats rose a foot in the air and fell back under a cloud of dust.

The whistle blew three times for the all-clear. The shovel returned, casually flung the mats aside, and started scooping up the loosened rock.

Even when blasting has been done, it is often necessary to break the rock into pieces small enough to be hauled away. This is where the *jackhammer* takes over. A jackhammer is a jumping-jack tool commonly seen battering holes in sidewalks or roads. Compressed air from a noisy engine is channeled through a hose into the body of the hammer. The air drives a weight downward with tremendous pressure against the jackhammer's pointed bit, forcing it into the ground. The weight is returned to its starting point by a powerful spring and then is slapped down again by the compressed air. This process is repeated more than 200 times a minute, causing the jackhammer to sound and work like a giant woodpecker. Big rocks soon split into manageable pieces.

Once all the rock is loosened, it must be scooped up. One of the machines used for scooping is the *back-hoe digger,* which looks like a small power shovel except that the shovel itself seems to be mounted backward on the end of a jointed arm. The operator sits in his glass-enclosed cab controlling the steel-toothed scoop, which jerks its load toward the machine like a man with a hoe.

As soon as these diggers started to punch and claw the

The camera has pulled back so that the whole excavation is visible. Along 49th Street at the left the steel superstructure is up to ground level. The wall of the "China Fair" and the scaffolding below it are in the far corner.

earth, a fleet of forty trucks started driving up to carry away the rubble. The trucking company's assignment was to keep the trucks coming so that the shovels and loaders, which rented for several hundred dollars a day, would not lie idle.

Trucks must also have a place to dump their loads. The big Macks and Internationals took the dirt and rock from the Exxon excavation through a tunnel under the Hudson River to a swamp in New Jersey, where the owner wanted "clean fill" (unclean fill would be trash or garbage).

Since the trucks had to go along city streets, any spills would be dangerous and messy. Workers had to be extra careful that the loads were securely packed. So the big, clumsy loaders and shovels tamped down the rocks and earth into the body of the trucks. Finally, as the loaded trucks lumbered away, a man with a crude metal hook reached up and scraped away the last loose debris.

After six months of blasting, scooping, loading, and carrying, the great digging job was done. The floor and walls of the hole were squared off roughly, ready for sealing and finishing. The excavators and their huge, noisy machines were hauled out of the pit and moved away to some other site to dig again.

HOW A SKYSCRAPER STANDS

The first task of any builder is to deal with gravity. This powerful force pulls everything toward the center of the earth and seems to resist any effort to build upward. But over the centuries builders have found ways to use gravity in keeping a building up.

Since gravity pulls straight down, the first rule of construction is to build straight up. If a building is properly supported, gravity actually helps hold it in place and keeps it from falling down. In order to use gravity, a building must also be firmly connected to the earth. Its foundation must not be able to sink or shift in any way.

These principles have been understood for centuries. Yet skyscrapers weren't possible until the last hundred years. The reason is that the necessary materials and building techniques were not developed until the late 1800s. In order to understand how the Exxon Building went up, we must first take a look at these new developments.

Most large buildings were once made of stone or brick. In such buildings, the walls usually supported all of the weight. Load-bearing walls must be built very solidly. It is impossible to put very many windows in them, and they must be very thick at the bottom. Builders long ago learned that buildings of this type could not be more than six or seven stories tall. To go much higher, the walls at the bottom would have to be so thick that there would be little space left inside. And the weight of the walls would be so great that the lowest tiers might crack and crumble.

At the same time, everyone knew that smaller buildings with no more than three or four stories could be made of wood. A simple framework of wooden columns and beams could support the weight of the structure, and the walls themselves would not have to carry the whole load. But wood is not strong enough for a large building.

About a hundred years ago, builders in large cities began to feel the need for structures with more than six or seven floors. But they had to find some material that could be used to strengthen the support. At first they tried reinforcing the walls with iron. Iron was strong, but experience taught that it was either too brittle or too soft to be very effective. Concrete was another promising material for high structures. But by itself, it was so brittle that it could not be used without metal reinforcement.

The building of modern skyscrapers did not really become possible until the strong, hard alloy of iron called steel began to be made in large quantities. Of course, flat steel walls were not the answer either. They would be as heavy as stone and more expensive. But steel beams can be used as a frame that will support many times more weight than a wood frame.

As builders learned to make better and better steel beams, it became possible to build 50 or 100 stories into the sky. The new beams became known as *structural steel.* They were not solid, but were specially shaped into the form of an I or an H, making them far lighter than a solid beam and nearly as strong.

When structural steel came into wide use, huge, sky-scraping towers began to loom on the skylines of major cities. The Woolworth Building in New York reached nearly 800 feet in 1912. Then came the Chrysler Building and the Empire

One of the first true skyscrapers, the Reliance Building in Chicago, rises next to five-story stone structures. Its most important ingredient: steel.

State, each over 1,000 feet high. Today, most major cities in the world have at least one building that is over 500 feet high. And nearly all of these buildings have been built by the same method.

First a steel skeleton is put up. This skeleton will support the whole weight of the building. Putting on the walls becomes just a finishing touch. Composed of glass and other fairly light materials, the walls are just the skin of the building. They support no weight and are used basically to keep the weather out and the people in. Walls of this kind are called *curtain walls.* Since they carry no weight, they might even be "hung" from top to bottom like curtains.

Other new developments were necessary before skyscrapers would become practical. Without workable elevators, skyscraper occupants would not be able to get back and forth to their lofty spaces. New, powerful pumps would be needed to bring up water for washing, drinking, and flushing. And some rapid form of communication was necessary unless workers planned to have a huge staff of messengers to carry information from one office or floor to another.

By 1900, electricity was available to power the elevators and pumps, and telephones (which also rely on electricity) had provided a rapid means of communication. In recent years, electricity has become an important source of energy for building skyscrapers as well as for working in them.

When architects and engineers sit down to plan a skyscraper, they have to determine exactly what forces might upset its stability. There are four kinds of load they have to consider. The first load is the *dead weight*—the total weight of the building when it is finished. The second is the *live load*—the weight of the people and equipment inside the building.

The third kind of load is *wind load.* The finished building will have a huge surface area exposed to the wind. Wind does not pull down on the building like gravity, but pushes sideways, trying to topple the big structure. The force of wind pressure can be understood when you discover how easy it is to blow down a stack of six or eight wooden blocks. A building with one story stacked loosely on top of another would topple in a windstorm almost as easily as the blocks. Modern skyscrapers are specially braced to resist the pressure of the wind. Surprisingly, they are designed to be flexible, so that in a high wind they may even sway a few inches.

The fourth kind of load is the stress that might be caused by an *earthquake*. This force would push or pull at the part of the building below ground level and shake the upper parts loose. No serious earthquakes have occurred in New York in living memory, but in California and other places where earthquakes are more frequent, special precautions must be taken in designing a tall building.

Once the extent of these four loads has been determined, the architects and engineers can decide how large the foundation supports have to be, and how large and strong the steel beams must be at any point in the building. They can then make provisions for special stresses, such as extra bracing against high winds.

Like most modern skyscrapers, the Exxon Building was designed to have a *core* section running up through its middle. This core is the real tower of strength. It is made of very strong, heavy steel columns and beams, and its sides are braced with diagonal steel members to make it even stiffer. This cross-bracing is important in resisting the pressure of the wind. The core columns support most of the building's weight. This means that the outer skeleton—the columns and beams near the outer walls—may be built of lighter steel. The core usually contains the life lines of the building—elevators, major water and sewage pipes, and air-conditioning and heating ducts. A typical floorplan for the Exxon Building shows the position of the core clearly.

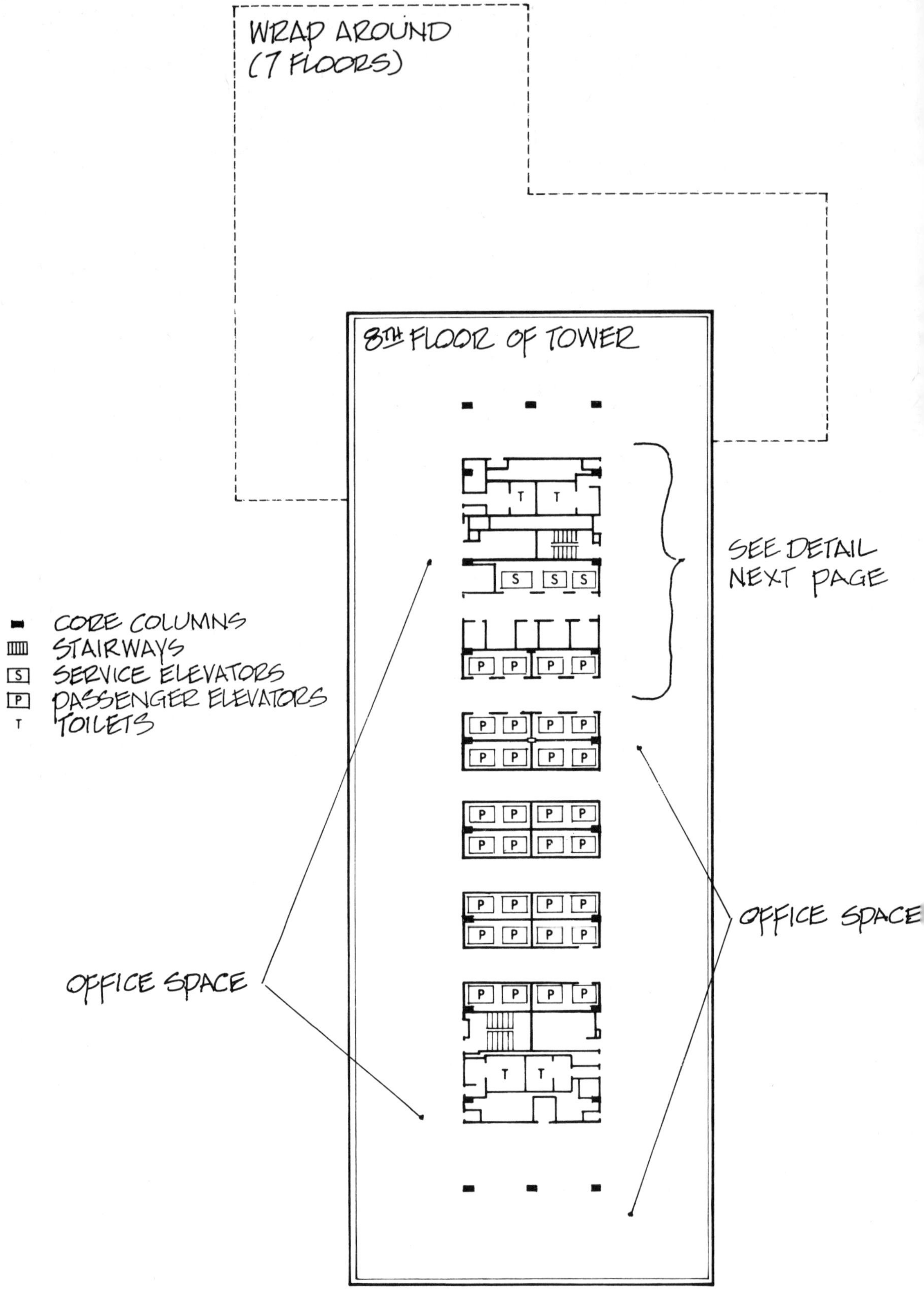
WRAP AROUND
(7 FLOORS)
8TH FLOOR OF TOWER
T
T
S
S
S
P
SEE DETAIL
NEXT PAGE
CORE COLUMNS
STAIRWAYS
SERVICE ELEVATORS
PASSENGER ELEVATORS
TOILETS
OFFICE SPACE
OFFICE SPACE

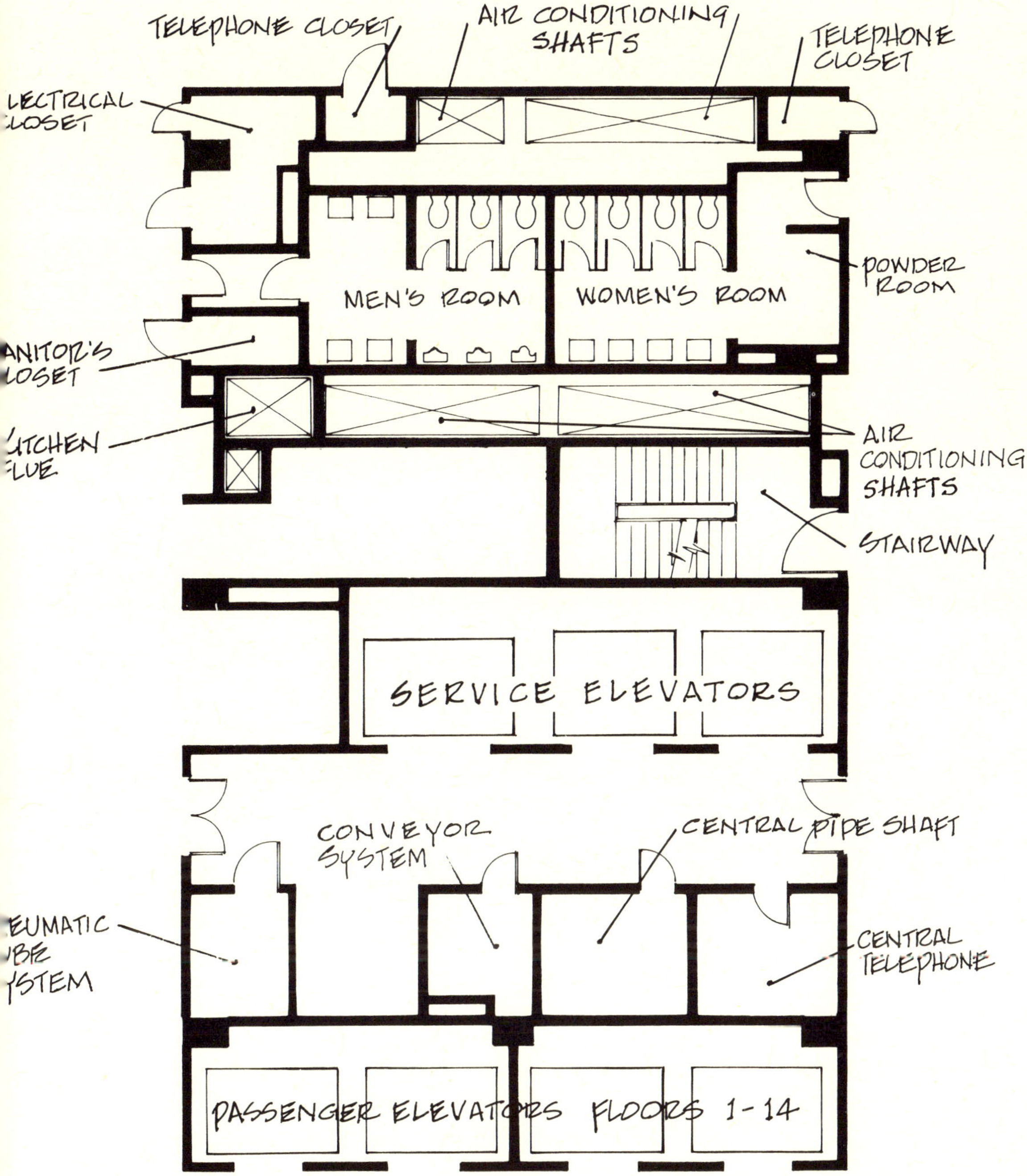

In the floor plan of the Exxon Building's eighth floor (left), the strongly built core—containing elevators, plumbing and other heavy equipment—is surrounded by office space which will be built of lighter materials. Above is a detail of one portion of the core, showing the location of equipment and service areas. The main core columns are shown in black.

Steel makes it possible to leave large open spaces—widely spaced columns are strong enough to carry the weight of the floors above.

Thus we see that a skyscraper is the result of combining ancient building principles with modern materials and techniques. The designers have called for a skeleton of structural steel strong enough to carry thousands of tons of weight and withstand the push of hurricane winds, yet light by comparison to earlier types of buildings. When it has been enclosed in attractive and relatively thin curtain walls, the huge building will provide a safe and comfortable home for thousands of office workers.

In order to make such a plan a reality, the building still requires the craft and skill of thousands of workmen. As we have seen, the first of these have already hollowed out the excavation. Now others must plant the foundations on which the skyscraper will rest.

THE FOOTINGS

In some ways the Exxon Building is very light. But by most standards, the building is immensely heavy. Its dead weight is estimated at about 225,000 tons, or more than 450 million pounds.

One of the builders' first concerns was to find a way to make the ground support this tremendous weight. If the bottom columns rested directly on the rock at the bottom of the excavation, they would sink deeper and deeper into the rock as their load increased. Thus it was necessary to find a way to distribute each column's load over a wide area, and to use many columns to share the total weight.

At the same time, the designers and builders had to be sure that the *footing*—or base—on which each column would stand was immovable and that the building would rise absolutely vertically. A few inches of tilt could cause serious structural damage to the building.

A large portion of the island of Manhattan is almost perfectly suited for the construction of skyscrapers because its bedrock is strong enough to bear the weight of large buildings. In other parts of the country, bedrock may be buried under a hundred or more feet of sand or clay. But in this central part of New York City it is so close to the surface that it sticks up in plain view in some park areas. This solid, almost immovable rock made the digging hard for the excavators, but it provided a firm support for the building.

A hundred and eight footings—one for each column—now had to be prepared so that all of the columns could be set at the proper level. Leveling the bedrock itself would have been very difficult. Fortunately, the builders had a more efficient way to make the footings. They used a kind of liquid rock: concrete.

Raw concrete is a wet, sticky mixture of cement, sand, gravel, and water. It can be smoothed glassy and flat for floors or poured into molds or forms for columns and walls. It becomes solid in a few hours and hardens in a few days. In a few weeks it is as solid as many natural rocks. The great advantage of concrete is that it will take the exact shape desired.

This footing is being poured directly from the concrete truck. The men inside the form are standing on a temporary wooden platform. The man in the center of the form is using a vibrating machine to "shake down" the concrete.

The excavators had left the bottom of the excavation roughly level. Now the foundation workers blasted holes in the bottom of the excavation—one hole for each footing. Then they built wooden forms in the holes to receive the wet concrete.

At about ten o'clock on a raw January morning in 1969, the first concrete truck arrived at the Exxon Building's excavation. The great drum was slowly turning to mix the concrete as the truck backed into position near the form for the first footing. A crane swung ponderously around, dangling a giant funnel-shaped bucket, or hopper, with a hatch on top and a closed spout at the bottom. The hopper dropped precisely into position behind the truck, and the mix splashed down into the hopper.

When the hopper was full, the crane swung it up and around. It dangled in the air a few feet above the wooden form, where three men were waiting. One of them opened the spout and directed the flow of concrete into the form. The concrete soon covered the maze of steel reinforcing rods. Sticking up through the concrete were metal connections which would soon be attached to the first column. Then a man brought a vibrating machine that looked like a garden hose with a long steel tube on the end. He put the tube into the mush and "shook" the raw concrete until it filled every chink in the form and was packed solidly around the reinforcing rods. When the concrete reached the top of the form, the crewmen scraped the excess concrete off the top and flattened the surface around the connections.

The first footings were poured while the excavation work was still going on. Not all of them could be finished at once. Instead of wasting time, the concrete workers started to pour parts of the basement walls and basement floor. Soon, various crews would be working on footings, walls, floors, and eventually the erection of steel—all at the same time.

Three months later, on a blustery April day, a monstrous construction arrived on a huge trailer truck. It looked something like a large raft made of steel beams. It would be the first piece of structural steel to go into the building. The occasion was so important that officials representing the owners, architects, and contractors were on hand to have their pictures taken.

The steel structure was called a *grillage.* It would fit between the concrete footing and the first giant steel column, which would help support the core of the building. The grillage weighed 40 tons (80,000 pounds) and would spread the load of the single column over an area 12 feet square. It would support a weight of 6,500 tons (13 million pounds) without bending or shifting.

The crew of ironworkers that day consisted of eight men. Together with the operator of an immense Manitowoc crane, they lifted the grillage "all by themselves." The first four attached stout crane cables to the grillage on the trailer and supervised as the big crane raised it slowly off the truck. The second four guided the grillage slowly downward until it settled perfectly onto the concrete footing. Then it was bolted down.

Twelve of the huge grillages were to be set around the center of the excavation, to support the massive central core of the building. Lighter steel plates called *billet plates* would rest on the other footings to support the lighter columns that would carry the outer part of the building.

The concrete footings were within inches of the same level, but that was not close enough. The grillages and billet plates were now adjusted so that each column would sit precisely at the level planned.

This operation was less spectacular to watch than the previous one, but it was particularly important. The key workers were the surveyors with their little leveling telescopes, called *transits.* Working from an established level, a surveyor aimed

The first grillage—a huge raft-like steel assembly weighing 40 tons—is lowered onto the finished footing at the right.

his transit at a pole held straight up and down on a billet plate. He leveled the transit carefully, then peered through the eyepiece at the pole. If the target disk on the pole appeared half an inch below the cross hairs, the plate had to be raised by half an inch. To raise a billet plate, a crane lifted it while workers inserted stainless steel wedges underneath.

The surveyor sighted again. Final adjustments were made, and the plate was okayed for level and bolted down. Then another crew came around and forced special high-strength cement *grout* into the narrow space between the billet plate and the footing.

When the grout hardened, the plates were permanently fixed in position. The solid "feet" of the building were anchored in place. Now the first columns could be raised.

The grillage is in place. It will now be leveled, high-density concrete will be forced between the steel members, and the pit will be filled with concrete.

CRANES AND DERRICKS

During the earliest stages of the digging, the busiest lifting machines were the power shovels and loaders. But now, as the concrete footings were hardening and the heavy steel began to arrive, a new kind of lifter was needed. It was the Manitowoc crane, which would tower 150 feet above the bottom of the excavation and be able to lift grillages and steel columns that weighed hundreds of thousands of pounds.

The huge machine arrived at the site in pieces on big flatbed trailers. The body of the crane, on which the operator's cab was mounted, was on one trailer. Other pieces, including the disassembled sections of the *boom* and the massive *counterweight,* were on other trailers.

The first problem was to "walk" the big crane body from the street level down the muddy ramp into the excavation. The crane was not built for moving very far or very fast; in fact, it looked like a factory building on caterpillar treads. The crane operator had a touchy job to do, but he didn't seem worried.

The diesel engine was already started when the operator swung himself up onto one of the giant treads and then still higher into the control cab. He patiently checked out every control before he began his descent into the excavation. Surrounding him in the cab were thirteen control levers and at his

feet were several pedals. These levers and pedals would allow him to control the crane once lifting began, but most of them were useless for running the crane body down the ramp. Now he was mainly concerned about controlling the two caterpillar treads on which the crane body rested. The only way to steer was to slow or stop one tread while the other continued to push. It was like driving a tank.

Finally the operator decided everything was ready. He eased the crane into position at the top of the ramp and then, almost without pausing, headed down. The machine rocked forward and started its slippery walk.

The ramp was so steep that the crane threatened to coast and skid down faster than it could safely go. Mud and slippery rock under the treads made it slip and slide almost uncontrollably. The machine slewed sideways several times, as onlookers watched nervously. Once or twice the operator braked one of the treads suddenly. This produced heaving, skidding turns that might have caused the crane to slide off the edge of the ramp.

But the operator kept control. Fifteen minutes after he started down, he rolled his machine forward onto level ground. Trucks carrying the boom sections and counterweight followed, and workers started putting the crane together.

The sections of the *boom* were lifted off the truck and placed on the ground end to end by a smaller crane. This metal framework is the huge "arm" over which the crane's lifting cables would pass. It would look almost cobweb-fine when standing 150 feet tall, but on the ground it looked much thicker and stronger. Each boxlike section took up as much room as a small truck and the main tubular steel pieces were as thick as a man's thigh. Workers bolted the sections together on the ground and attached the lower end of the boom to the front pivot point of the crane body. On the upper end of the crane they attached a smaller boom, called a whip boom.

The Manitiwoc crane at work in the excavation. Behind it is the ramp it came down.

Next, the huge *counterweight* was hoisted and bolted to the back of the crane body. The crane is really a kind of balance. The objects it lifts are in front of the body, and the counterweight is in back. These two weights must balance each other or the crane will tip over.

Now four main sets of cables were unreeled from the body of the crane (where they were connected to a system of winches and pulleys) and attached to the boom. The first set would be used to lift the boom itself. The second set would control the angle of the smaller whip boom extending from its tip. Once the boom was completely assembled and the counterweight was in place, the crane operator climbed into the cab again and started his engine. Then he began to pull the boom cables taut. As they tightened, the force squeezed the boom sections together, making them stronger and more rigid. The crane reeled in more cable, and the huge boom lifted slowly off the ground. It rose majestically until it pointed straight up. The whip boom control cable automatically kept the fingerlike extension at a constant angle.

The other two sets of cables, which had been attached to the boom before it was raised, were the hoisting cables. The *main hoist cable* (or lifting cable) was passed over a pulley at the very tip of the main boom, and the whip cable was passed over a pulley at the tip of the whip boom. The whip cable would be used to lift light objects to a greater height. The main hoist cable would do the heavy work. It was 1 $\frac{1}{8}$ inches in diameter—about as thick as a shower curtain rod. A steel ball was attached to the lifting end of the cable to keep it taut when the crane was not carrying a load. Below the ball was a big steel hook.

The other end of the cable entered the cab through the roof and was wrapped around a big steel drum. When the operator pulled a lever, the drum turned and the cable wound on very neatly and evenly without being guided—the weight of

The crane body is bigger than it seems—compare it to the man working on its treads at right.

the cable and its load was enough. When the first layer of cable had been completely wound on the drum, a second layer was automatically started, and so on.

Soon after the crane was fully assembled, the operator began work by lifting the first grillage into position. Within a few days he was unloading whole truckloads of steel from street level. The tall boom would swing slowly into position as the operator rotated the cab. Then he would unreel the main hoist cable until the hook hung a few feet over the truck. A stack of beams had already been looped with short slings of cable by a crew of ironworkers. Now they put the loose ends of each loop over the crane's hook.

When all was ready, the operator received a hand sign from the signalman to lift. Carefully, gently, the hook rose until the cable loops formed a tight sling and lifted the bundle a few inches off the truck. If the bundle started to tilt, the signalman would give orders to lower it again so that workmen could adjust the cables. Then, after another test, the whole stack rose, guided by a man on the ground who held a rope attached to one of the beams. The heavy load swung up and around and then lowered gently to the ground at the proper place.

During the actual lifting, the signalman gives the orders. Although the crane operator can usually see what he is lifting, in some situations his view may be blocked. But the signalman on the ground must always be in a position to see the load from the moment it is lifted until it is finally put in place. The hand signals hc uscs can also be used by the ironworkers to communicate above the noise of the construction. They can ask the time of day or even carry on a simple conversation.

All that spring and summer, the great Manitowoc crane could be seen lifting and setting down huge loads of steel. In addition to unloading trucks, the crane also lifted the heavier columns and beams into position. This was even more delicate

The delicate-looking boom of another crane lifts supplies on another New York building site.

work, since each piece had to be set down exactly where it would be fastened into position. The piece would hang in position from the hoist cable while ironworkers fastened it into place with temporary bolts.

Finally, after four and a half months, the Manitowoc's job was done. So much steel and concrete had been put together that there was no more space in which the 15-story boom could maneuver. It was taken apart, much as it had been put together, and the crane body walked the treacherous ramp once more—this time uphill to street level.

But the lifting jobs had hardly begun. Before the Manitowoc was dismantled, three other lifting devices called *derricks* were set up on the ground floor. The derrick works from a fixed position. It has a vertical mast, steadied by cables, and a movable boom. Derricks are more limited than cranes, but they take up less room. Their operators often do not even see

The derricks jutting out from the Exxon Building and the outside hoists running up its side are two more important lifting devices.

what they are lifting. They sit in their control boxes all day, raising and lowering things in response to bell signals—one ring for "up," two for "down," and three for "stop."

A derrick needs both its mast and its boom for one special reason. As the building goes up around it, the derrick is soon too low to be useful any more. When this happens, it is partly dismantled and the boom is lifted by the mast cables to the next working floor. Then the boom is set up as a new mast and its cables reach down to lift the lower mast up to the higher level. By this bootstrap method, derricks lift themselves floor by floor right up to the top.

Even after the Exxon Building had been completely enclosed, one of the derricks remained on one of the top floors, its boom thrusting out beyond the sheer wall. A hole was left in the wall below the derrick so that last supplies could be pulled into the building.

After the building skeleton had been raised a few floors above ground level, workers put together still another lifting device. It is called an *outside hoist* and is really a rough elevator running inside a scaffolding that is attached to the outside of the building. The scaffolding is always several floors taller than the building skeleton, and as the building goes up, the scaffolding grows, too, giving the hoist additional steps as floors are added.

The operator of the outside hoist works at ground level. He sits in a booth and runs the hoist engine in response to bell signals. The hoist cable passes right beside him and is marked to show him which level the cage (or car) is on. When he gets the signal to lift to the 40th floor, he watches the marks as they go by. They tell him when to stop.

But the outside hoist gets us well ahead of our story. Actually, the grillages and billet plates have only just been put into position and they await the placement of the very first steel columns, the beginning of the skyscraper's skeleton.

FIRST COLUMNS

Now it is April 1969. It has been thirteen months since the first workers arrived on the site to tear down the old buildings. Eight months ago the excavation was begun. The first grillage has recently been set in place and the activity in the excavation is infinitely varied and infinitely noisy.

On the west end of the hole, the smoothing of one rocky wall is not yet complete. Pneumatic drills (operated by compressed air) are still preparing holes for explosives, and blasting still takes place several times a day.

In the middle of the site, wooden forms in the rocky floor are awaiting the arrival of concrete. The last footings must still be poured. Nearby, another grillage is being positioned on a completed footing, and billet plates are still being placed on the outer footings.

To the east, near the Avenue of the Americas, things are a little farther along. A concrete floor has been poured around the positioned billet plates and is already half-invisible beneath stacks of materials and rubble. Along the north edge, forms are in place for the first basement wall and one crane is standing by to help fill the forms with concrete.

In the midst of all the other activity, the real construction is about to begin. The workers have prepared the foundations for the huge building. Now several grillages are ready to receive the first mammoth columns of steel—the ironworkers begin to build upward.

The Manitowoc crane is needed again. The first columns to go up will be the heaviest ones used. They will share the enormous weight of the core of the building. They are two stories long and weigh more than 2,100 pounds per foot of length—nearly 25 tons in all. All vertical members, or columns, will be H-sections. When seen from the end, their shape is roughly an H. But these first columns must support

The first column is ready to be placed on its grillage. Held aloft by the crane, it is steadied by two workmen who hang on to the flanges that will attach the column to the grillage.

Workmen push and strain as the column is lowered the last few inches onto the grillage.

so much weight that except for shallow depressions on two sides, they are solid shafts of steel (see photo on preceding page). Their sides are covered with an off-red paint that protects them against rust, but their bottoms shine like silver. They have been machined so that they will sit squarely on the grillages.

A cable sling is wrapped around the first column. The great Manitowoc crane crawls close and then lowers its main hoist cable so that the sling can be thrown over the big hook. The signalman gives his thumbs-up. The giant column stirs, and the crane's diesel engine sings a deeper note as more power is applied. Gently the column rises until it is almost vertical. Then it lifts free. The crane body and boom turn ever so slightly and the column swings toward its position, guided by a man who holds a rope tied to the bottom of the column.

Slowly the 25-ton steel piece drifts right over the grillage to which it will be fastened. Four connections stick out of the grillage top. Their bottom ends are buried deep in the concrete footing. The crewmen grab the column and twist it until the connections, or *flanges*, on the bottom of the column stand right over the connections on the grillage. Then the signalman turns his thumbs downward and the great column sinks gently into place. The crewmen run bolts through the flanges, spin nuts down on them and tighten the nuts with large wrenches until they are snug. Finally, other bolts and angles are used to fasten the column into exact position.

Now the crane relaxes its pull and lets the noose slip down the column. A crewman releases the great hook and the crane swings away, leaving the first column standing free.

As soon as a second column is installed nearby, things speed up. A lighter crane crawls in to lift the first crossbeam, or *girder*, while the Manitowoc is helping to position a third column nearby. Ironworkers shinny halfway up the first two columns to fasten the first girder. Guided by means of ropes tied to both ends, the girder floats between the two columns. One end slips into position and the worker on the column drops his tapering *driftpin* into the matched holes of the girder and the column's flange. The driftpin will hold the girder in place until its other end is in position.

Fitting the second end is not so easy. The second column —which, like the others, has not yet been permanently fixed at its base—is tilting slightly inward. The worker who shinnied up that column now steps out onto the floating girder and pushes against the column. It moves slightly and the beam slips partway into the slot. The worker climbs back onto the column and hits the edge of the girder with his sledge hammer. After a few ringing blows, it yields and slips completely into place. He drops his driftpin into a hole and then the workers at both ends proceed to make the girder fast by inserting temporary two-inch construction bolts through the holes and tightening nuts down onto them. When they finish, the first two columns and the girder form a large letter H.

Now the steel-raising pace intensifies. The third column has been positioned and work is proceeding on the fourth, completing a square. When girders have been put up connecting each of the columns, the first square area, or *bay*, has been finished. Now the crewmen climb to the tops of the columns and get ready to connect them with four more girders around the top. Soon after the columns have been connected by heavy girders, other crews move in to lay a row of lighter I-beams, called *filler beams*, from one main girder to another. They use the samc mcthods, and one of the smaller cranes does the lifting. When the filler beams are in place, a temporary floor of heavy wooden planks is laid across them for the workers' convenience. The area beneath the floor is thrown into shadow. Within weeks, the whole bottom of the excavation will be in darkness.

For three weeks, steel is raised in this manner. A basic

Within weeks the skeleton has reached street level. Wood planks make temporary floors, and the bottom of the excavation is thrown into darkness.

crew of ironworkers is small—only five men—but each crew is fast and sure. Now the skeleton has risen several stories on the east end of the excavation. The surrounding activity continues, too, and sometimes two different kinds of crews must cooperate to get a job done.

On the north edge of the site, concrete workers have put up a scaffold next to a concrete wall which they are waterproofing. Then the ironworkers come to attach one final girder to a column that stands behind the scaffold. After asking the concrete workers to move out of the way, they try to swing the girder into place with the crane. But the girder strikes a vertical pipe on the scaffolding and will not move into place. For a few minutes the ironworkers struggle with the scaffold, trying to budge it only an inch or so to make room for the girder. But it doesn't work.

Reluctantly the crew faces the fact: the scaffolding will have to be moved. Quickly they set to work. They pull the wooden planks aside and, with the help of the crane, lift away a whole section of the scaffold. Then they swing their girder into position, fasten it, and put the scaffold back in place. After this unexpected break, the concrete workers climb the scaffold again and get on with their job.

Overcoming these and other minor problems, the erection crews worked surprisingly fast. By June 20, only two and a half months after the first column went up, the steel skeleton filled the excavation completely. Temporary cables of steel crisscrossed the skeleton to hold the columns vertical and the angles square.

As the erection crews continue to set new members into place, other crews of ironworkers follow along, reworking each connection. After removing the temporary construction bolts, they insert permanent ones of hardened steel and drive the nuts home with pneumatic (compressed-air) wrenches. Before this finishing crew leaves a connection, an inspector checks it for bolt tightness, level, verticality, and angle. He is responsible for seeing that the building really does go straight up.

The next crew to arrive removes the temporary wooden floors and replaces them with flat metal pans which rest on top of the light floor beams. The wide, shallow pans, called "decking," are permanent forms for the concrete floors. Their bottoms contain open channels in which electrical and telephone wires will be installed. Then these channels will be covered with steel plates and the bottoms of the decking will form the rough "ceiling" for the floor below.

As these pans are put into place, the last daylight is blocked from the lower levels. What used to be a hole in the ground is being turned into a basement several stories deep. As the building goes up, workers have more and more need for light for indoor work, electrical power for lifts, and water and sewage connections. We will soon see how these connections are made.

But first we must make a short side trip. The skyscraper is being made possible by the enormous steel members that are now being set in place. Where did they come from and how did they arrive at the site?

The area in front of the building was excavated to make room for a subway concourse, storage and utilities. Special wood flooring at street level makes it useful during construction. Later there will be a pool, shade trees and benches (see model, page 23).

THE BIRTH OF A COLUMN

The magnificent steel beam that began the Exxon Building's skeleton, like all the steel members in the building, was created by the Bethlehem Steel Company of Bethlehem, Pennsylvania. The story began with a carload of brownish iron ore which had been buried in the earth. Soon the beams that came from it would be buried, too—deep inside the skyscraper. But for a short period in between, the material had a violent existence as it was turned from ore to iron to steel.

After the ore arrived at the steel mill in a railroad gondola car, it was graded, sorted, and finally dumped into the top end of a huge blast furnace, along with measured portions of limestone and *coke* (coal that has been baked to burn out its impurities).

As the furnace roared mightily, the burning coke released gases that helped reduce the ore to molten (liquid) iron. At the same time, the limestone helped convert the impurities in the ore to *slag*, which floated to the top of the liquid and was skimmed away. More impurities were blasted out in the form of smoke. A complicated process collected tons of airborne impurities from every batch.

Finally, after five or six hours of this violent treatment, a shining stream of molten pig iron was drained from the bottom of the furnace into a long tubular railroad car that was called a submarine and looked like one. Sloshing around in this monstrous black vehicle, the molten iron looked like melted orange sherbet—at a temperature of 2,900 degrees.

The submarine moved the metal to another building that contained two huge Basic Oxygen furnaces, where the iron would be turned into steel. By the old open-hearth method this process would have taken eight hours. But by the new Basic Oxygen method, it would take only forty-five minutes to make a batch, or "heat," of steel—approximately 270 tons. The furnaces themselves were enclosed in vast corrugated-iron sheds, side by side. They stood three stories high and looked like giant pear-shaped earthenware pots. In fact, their linings were a form of earthenware. A container made of almost any other material could not hold the molten steel without melting itself.

The "submarine," now standing near the furnace, was tilted and the molten iron was poured into a huge ladle which lifted it and poured it into the top of the furnace. Also added were bales of scrap metal—some of it crushed food cans that had been collected by conservationists. Small amounts of other metals, called alloys, were poured in, too.

The furnace cooked the ingredients until they reached the right temperature. Then pure oxygen was blasted into the mixture under high pressure. The oxygen helped burn out the last impurities in the metal and in a few minutes the iron had become steel. It would be purer, stronger, and more workable than the pig iron that had been poured into the furnace earlier.

Forty-five minutes after the ingredients had gone into the furnace, the steel was ready to be tested and moved. The huge pot was tilted until its roaring, sun-bright mouth was visible. Workers wearing protective woolen smocks and dark goggles dipped into it and removed samples that looked like Roman candles. The samples would be allowed to cool and then they would be broken and tested, first by the workers at the furnace

Molten iron is poured into a basic oxygen furnace at a Bethlehem plant.

and later in a laboratory. The workmen also took the temperature of the white-hot liquid in the furnace with a rod that transmitted its findings to a computer cage nearby.

When everything tested out properly, an overhead crane with two great hooks moved in and picked up a huge ladle. A measured amount of the molten steel was poured from the furnace into the ladle. Then the ladle was moved until it hung over a row of black, barrel-shaped molds known as ingot molds. A worker pulled a handle which opened a spout in the bottom of the ladle, and once again the gloom was lighted by the glowing stream of metal. The steel would begin to cool in the molds while it was awaiting the next step.

The steelmaking process was complete, but the metal still had to be fashioned into a useful shape. The shaping of structural steel—the big beams that go into skyscrapers—is done in a rolling mill. While the steel is still hot and soft, it is forced through a series of mills which gradually squeeze it into the desired shape. The actual squeezing is done by super-hard steel "rolls," from which the rolling mills get their name.

When the steel in an ingot mold had cooled enough to become solid on the outside (although it was still liquid in the middle), the mold was removed. Then the barrel-shaped piece of steel, now called an *ingot*, was moved to a soaking pit, where it was "soaked" in heat until its temperature was the same all the way through. It was now almost white-hot and soft enough to be milled.

At this point, the steel is so hot that a man could scarcely get close to it without bursting into flame. It is moved from mill to mill on beds of rollers which are automatically controlled by men who sit in glass pulpits far above the clamor and heat of the mill itself. With the help of computers, they control each pass the steel makes through a mill. Then they send it to the next mill for further treatment.

The inside of the building where the beams were being milled looked something like a foggy twilight on a city street. The far end of the building was almost half a mile away and was practically invisible in the gloom. But closer at hand, the air was brightened by glowing bars of steel as they emerged from one mill and were sent on their way to another.

From the soaking pit, the yellow-hot ingot of steel was first run through a "blooming" mill which squeezed the barrel-shaped form into a bar about 25 feet long and one foot square. At the other end of the blooming mill was a reheating furnace where a visitor could observe the bars through a window as they were being heated for their run through the first rolling mill. The bars were almost too bright to look at, glowing with the color and intensity of a 100-watt light bulb.

Suddenly a set of rollers under one of the bars began to spin and the bar lurched forward into a channel whose movable walls clanged together to steer the bar toward the next mill. The bar nosed into the opening between two powerful rollers and disappeared. Almost immediately, the bar came back through the mill. Now it was smaller in diameter and slightly longer than it had been a moment ago. The handling walls slapped it again and guided it back through the mill. Each pass was tighter than the last, and by the seventh time through, the bar had a kind of dumbbell shape and had doubled its length. The mill was so powerful that the metal slapped back and forth as it was pulled through, like a fish swishing its tail.

Now it was ready for the next mill, but before it got there its front and tail ends were chopped off, since they might bend or fray and might not pass smoothly into the next mill.

The operators of the mills worked with complicated sets of instructions from the builders of the Exxon Building. Each column and beam was made to order for a special place in the new building. When the pieces were completely milled, they

A blooming mill begins shaping the glowing steel. The resulting "bloom" will eventually be milled into a steel beam.

would be cut to specified lengths and each would be marked with a code number that would tell the ironworkers at the building site where it was to go. Then the pieces would be sent to a machine shop where connecting flanges would be welded on and holes drilled to the builder's precise specifications.

The next mill was more complicated than the first. It had four rolls that gripped the piece of steel on all four sides. The top and bottom rolls squeezed the middle of the beam, called the web, flatter. On each side, the other two rolls kept the material from bulging sideways and squeezed the flanges—the two vertical parts of the H—into shape. The bar made three passes through this mill. When it came out, it was almost 200 feet long and was beginning to look like a finished H-beam. It had cooled to an autumnal red-yellow color. The last two mills simply refined the shape of the still lengthening beam.

Finally the completely milled piece of steel parked itself on a cooling bed where air could rise from beneath a grillwork and carry off heat. By this time it was a glowing cherry-red color. As it cooled it was slowly moved sideways, following other cooling pieces and making room for other hotter members that were coming from the mill. By the time it reached the edge of the cooling bed, the beam had darkened to a dull gray. The color spectacle was over.

The huge steel member, now some 240 feet long, went through a final mill which forced it into a perfectly straight shape. Then it was automatically rolled toward the buzz saw that would cut it into proper lengths. At its present length, the member would be too long to be moved by any truck or railroad and too heavy to be lifted by any crane. It slid forward until its front end butted against a stopping gate. Then the buzz saw, the size of a dining room table, dropped into place. With a frightful shriek and a fireworks display of sparks, it cut the member into beams of the proper length.

Now the beams were ready for marking and for receiving the required connecting flanges and holes. When they left the machine shop, they were ready for placement in the new building. A few days later, a flat-bed trailer truck arrived and carried a load of steel members to a storage area near the construction site. There they were stacked in the order they would be used.

After passing through a series of mills, a 240-foot length of structural steel emerges (bottom right). After it is cooled and cut to size, connecting flanges will be attached and holes drilled according to the builder's order.

Some light finished beams are ready for shipment to a building site.

Then, a day or two before they were to be set in place, another truck hauled them off to the construction site.

Making steel and rolling it into structural members are crucial steps in the construction of almost any skyscraper. Without this complicated industrial process, the giant towers would be impossible to build. But meanwhile, back at the construction site, many other jobs were being done.

MAKING CONNECTIONS

Long before the excavation was completed, the planners of the Exxon Building were making preparations to hook up to the supply systems of the city. Even in the early stages the building needed water, connection to the sewage system, gas, electricity, telephones, and even steam.

One of the most pressing needs was water. The water supply system of New York is a fascinating story in itself. It provides about 1 ½ billion gallons of water a day (about 140 gallons for every man, woman, and child in the city). The water comes from reservoirs in mountains and hills as much as 150 miles away and is carried to the city in huge aqueducts.

To deliver this small ocean to users, the city began early in the century to blast enormous tunnels out of solid bedrock hundreds of feet beneath the streets. One of these tunnels lies some 200 feet beneath the Avenue of the Americas, passing right beside the Exxon Building. The tunnel is engineered so that the water runs downhill from its sources far outside of the city. Vertical pipes situated at regular intervals along the tunnel bring some of the water to a system of pipes or "mains" nearer to street level. The water is under such pressure that if one of the pipes breaks, the water will spout dozens of feet into the air. Even if it is repaired quickly, the water may flood basements and subway stations for blocks around.

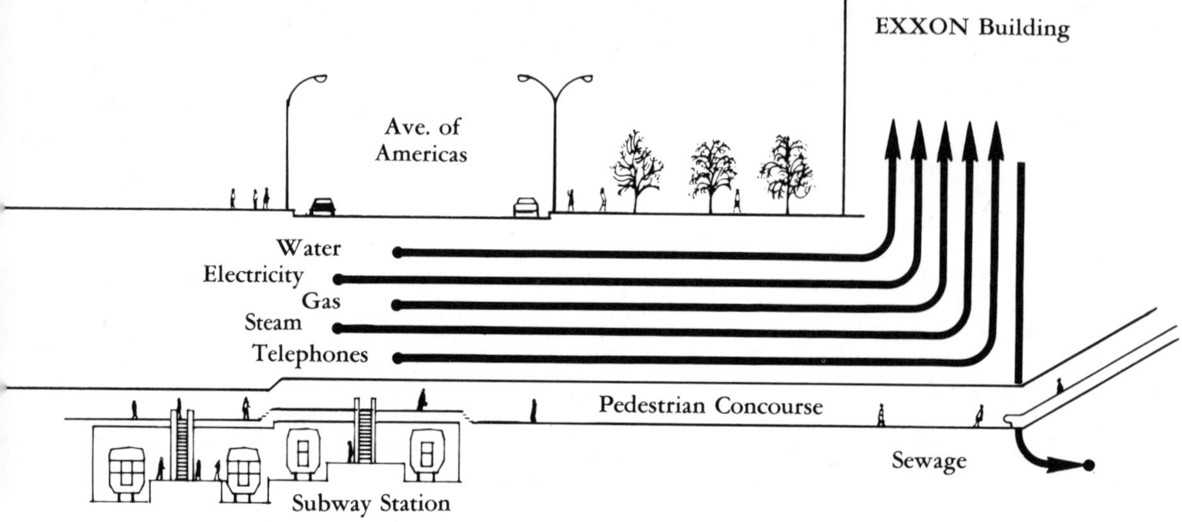

Huge quantities of water, electricity, gas and steam enter the building each day from under the city streets as shown in this schematic drawing. Even people enter and leave the building below ground level through the subway concourse.

These water mains ran under the street level at the corner of the Avenue of the Americas and 50th Street, right next to the Exxon Building. They were already providing water to many of the older buildings nearby. The builders should only have had to tap into a main to get water for the new structure. But it was not that easy.

The first problem was that the mains were in poor condition. Leaky pipes needed repairs before connections could be made. The main shutoff valve was so corroded that it was frozen in the "on" position. Water department workers tried for two weeks before they were able to close it. Finally the valve was closed and repairs made.

A second problem was that the pipe itself was about seventy years old and was made of iron. It was dangerously brittle. The inspectors were so worried about the pipe's condition that they decided there could be no blasting within 50 feet of it. This meant that one corner of the excavation had to be dug

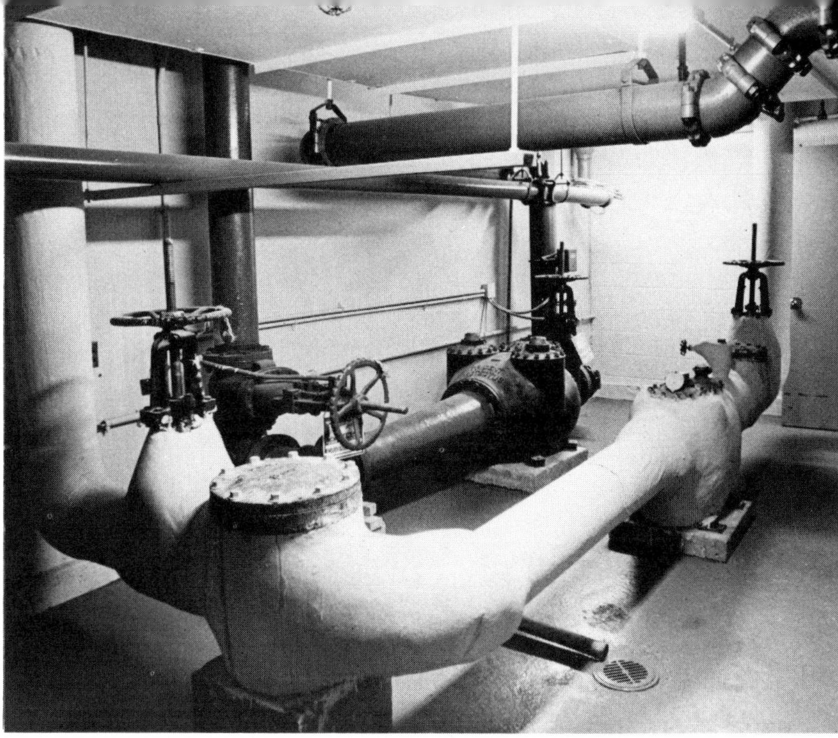

The light pipes in the foreground carry steam and are insulated to keep the steam from cooling. The dark pipes carry water.

with air hammers and other tools rather than with explosives.

After repairs were completed the connection for the new building was made by boring a hole into the wall of the pipe and inserting a waterproof collar which would accept a branch pipe to carry water to the construction site. It was a messy job. Even when the valve was shut off, water seeped out, so that workers had to wear hip boots and stand in muddy pools of water.

Other shutoff valves were built into the connecting line so that they could be turned off while the new building's plumbing was being installed or repaired. In the meantime, workers in the bottom of the excavation would have water for mixing concrete, drinking, and extinguishing any fires that might start.

Another connection was with New York's sewer system. This was needed early in the excavation to provide some place to pump rain water and underground water seepage that constantly threatened to make digging impossible. New York's sewer mains are pipes and tunnels big enough to walk in. They carry sewage from all parts of the city to treatment plants where it is decontaminated before it is dumped into the sea.

Hooking up the sewer system proved to be another headache. The architectural plans called for a direct walkway from the lower pedestrian level of the Exxon Building to the nearby subway station. But the great sewer pipe running down the Avenue of the Americas lay right in the way. It would have to be removed, but that would interrupt the pipeline.

The builders finally agreed with the city to build a new sewer main more than 600 feet long to connect with another part of the sewage system. Putting in the new line snarled traffic and took time, but in the end the walkway to the subway station was built and the city had a fine new sewer.

Later, when the building was almost finished, the builders installed a pair of special pumps called sewage ejectors, which force water and waste material out of the building and into the sewer.

Another concern was electricity. During the excavation phase, most of the machines used had their own power sources. Trucks, cranes, and loaders worked on diesel or gasoline engines, and even the jackhammers used compressed air from diesel compressors. But as construction began, electricity was needed to run tools and pumps and to provide light. Very soon it would also be used to power derricks and the outside hoist.

In New York, electricity is transported under the streets at thousands of volts of pressure in heavy copper cables sheathed in concrete. Before tapping into one of these cables, the builders set up transformers that would reduce the electricity to voltages suitable for normal use. Then they asked the power company to shut off power in the nearest segment of main cables so that the connection could be made.

Electrical workers clamped their branch cables to the main cables. Then they wrapped the connection with many layers of black tape and sheathed it in concrete to prevent a short circuit or a leak. When the power company turned the main power back on, electricity was available in the electrical "closet" and was immediately put to use. From this box, connections would be made to several other boxes that would eventually supply power to run elevators, lights, office machines, and other equipment on every floor of the finished building.

The Consolidated Edison Company, which supplies the electrical power, also supplies an interesting by-product—

The electricity room in the building's sub-basement—high voltages make it no place for a sight-seer.

steam. The steam was originally produced by the company to spin its huge generators to make electricity. Rather than waste the steam after using it once, Con Edison "dries" it and delivers it to large buildings through a network of pipes under the streets.

This steam is used to heat offices during the winter and to heat water. In the Exxon Building, it is also used to run huge air-conditioning turbines. The builders used the steam during the winter to heat sand and gravel for concrete, which does not mix properly if its ingredients are too cold.

Connections to underground gas mains were also made. The gas would be used in the finished building for cooking in the building's restaurants and cafeterias.

This left only the telephones to be attached to hook the building up completely with the outside world. At one time, New York's millions of telephone lines created an ugly forest of wooden poles and tangled cables. But by 1900 the wires were being buried under the street. Today they lie in tunnels encased in metal or plastic conduits as thick as a boy's leg. One tunnel may have as many as seventy-two of these conduits, and each conduit contains thousands of strands of thin copper wire, each wrapped in plastic insulation keyed by color. The different colors help telephone men pick out the right wires to make connections.

The telephone men began by attaching a branch conduit to the underground main and running it through a basement wall. The wires in the conduit were then extended to one master closet on each floor and then later divided again to reach out to four smaller closets. Although much of the preliminary wiring was done as the building went up, the job could not be finished until just before the office workers moved in. In all, about 5,000 telephones would be needed. They would provide communications between offices in the building as well as connection to the outside world.

Among the telephone cables were special hookups running to Western Union for telegraph messages, and to stock exchanges so that stock prices could be reported instantaneously to busy financial executives. By the time all the office workers had moved in, other special equipment might also be hooked up by telephone lines—such as teletype machines and even long-distance copying machines on which a letter or drawing in California could be instantly reproduced in New York.

The pipes and conduits and wires that connected the building to outside mains were its blood vessels and nerves. When they were all installed, the walls and ceilings were placed to conceal them. They were so well hidden that few office workers would ever realize how much their comfort and safety depended on them.

The underground concourse leads to the subway under the Avenue of the Americas and to other Rockefeller Center buildings.

CONCRETE

Next to steel, the most important material in a skyscraper is concrete. During the early stages of construction on the Exxon Building, hundreds of tons of mixed concrete were brought to the site in ready-mix trucks. The trucks had been loaded with the proper amounts of sand, gravel, cement, and water on the outskirts of town, and the mixing had been completed on the way to the site.

When the concrete was poured to form the basement walls and floors, an inspector was always on hand to check the mix for proper proportions and to take a sample of each load in a small container. The sample was allowed to harden for several weeks and then tested for strength and hardness. If it failed the test, the builders would have to tear down that particular part of the construction and build it again.

When the construction reached street level, a complete tanklike concrete mixing plant was built on the ground floor, where the materials could be mixed more cheaply. Now, instead of ready-mix trucks, dump trucks brought in the raw materials. These were fed into the top of a huge mixing tank, and mixed concrete was drawn off from a spout at the bottom as needed. Of course, the inspection process continued as before.

By mid-1970 the skeleton had reached above the 50th floor. Meanwhile, other workers followed behind, placing the

During early construction ready-mix concrete is used for basement walls.

filler beams that would hold the floors and installing the metal pans that would serve as forms for the concrete. Plumbers, electricians, and heating men were installing sleeves for conduits and pipes that had to be in place before the concrete floors were poured. Then came workers to pour the concrete.

One summer day a visitor came to watch the concrete men at work. The first 29 floors had already been poured. Now it was time to pour the 30th, even as ironworkers were completing the skeleton hundreds of feet above.

The scene was one of scurrying activity, and the principal actors were a group of odd-looking buglike trucks. The front of each truck was a big bin, like an oversized wheelbarrow. On the back was a small gasoline engine and a narrow saddle for the driver. One after another, empty trucks came dashing along temporary wooden pathways, zipping precariously around right-angle corners, on their way to a tall rectangular tank near the edge of the floor. (No walls had been put into place yet, so at the end of a floor there was a sheer drop of several hundred feet to street level.)

Once they reached the tank, the trucks nuzzled up to it like young kittens trying to suckle from their mother. The tank was a supply hopper. It was kept filled by a derrick that lifted the wet concrete from the mixer far below. As soon as an empty truck was in position, a worker opened a gate in the tank and sluiced a load of raw, gray concrete into the truck's big bin. Then the truck backed away, spun dizzily around, and dashed off down the plankway while another empty truck moved into its place.

The full truck headed for a spot where the broad metal pans had been prepared for the concrete. Suspended in the

Pouring concrete on the ground floor: the little bug-like truck empties it into forms prepared with a grid of reinforcing steel rods.

trays was a heavy mesh of steel reinforcing which would give further strength to the new concrete floor. Nearby stood four workmen leaning on their hoes and shovels, ready to spread the soupy mixture. They wore rubber boots and their skin was covered with grease. Concrete is a powerful alkaline substance, and workers may suffer from sores and allergies if they do not protect themselves against it.

The truck pulled up to the edge of the prepared pan area and stopped abruptly. The bin on its front tilted forward and spilled the mix into the pan. After the workers had scraped the last of the concrete out of the bin, it slammed back to its normal upright position. Then the truck did its dizzy spin and sped away just as another full truck arrived.

As soon as the first truck was unloaded this way, the workers began to spread the concrete. Although it was soupy enough to pour out of the truck bins, the men had to use shovels and hoes to make sure that the whole form was evenly filled. They poked and stirred it to make sure that it was firmly packed around the steel reinforcing, filling the form up to the steel retaining walls that showed the right level for the finished floor.

When the whole area had been filled, the spreaders moved on to another part of the floor and another team of workers arrived. Their job was to give the concrete a level (but still rough) surface. They zigzagged a long, straight-edged board across the surface, scraping off extra concrete where the level was too high and filling in craters where it was too low.

It took five or six hours before a third team could come in and finish the job. By that time, the concrete was quite stiff, so the workers removed the steel retaining walls between floor sections and filled the cracks. Then they swept smooth steel trowels back and forth over the floor. This job, called floating, caused any remaining water in the concrete to rise to the top, allowing the men to form the final shiny, sealed surface.

A workman smoothes the concrete with a long flat board mounted on a pole.

All day long the little trucks ran back and forth. Although the floor area was considerably larger than a football field, the crew managed to pour the whole thing in a single day. When the trucks were finished, they were driven to the edge of the floor, where a crane was waiting to lift them up to the next level. One by one, the trucks swung out over the sidewalk and up onto the 31st floor, where they would work the next day.

Several levels below the new floor, other workers, called masons, were using another type of concrete called mortar to build cinder-block walls. As we shall see, many of the walls inside a modern skyscraper are prefabricated. These light, movable partitions usually separate one office from another. Still, the masons were needed to build permanent walls to enclose the elevator shafts, the washrooms, and some hallways and to

"Floating" the concrete with a trowel.

build small enclosures called bays under outside windows for air-conditioning equipment.

Masonry is one of the oldest building arts known to man. Nowadays, masons often make walls of cement or cinder blocks rather than bricks, but the process is very much the same. The first row, or "course," of bricks is laid in a sticky bed of mortar at floor level. Mortar is also used between the bricks to hold them together and seal the wall. Then a layer of mortar is applied to the top of the first row, and a second row is laid down. The bricks are never placed exactly one on top of the other. Instead, they are staggered or alternated so that the gaps between bricks come at a different place in each row. This arrangement makes the wall stronger.

The big trick is to make the whole wall both straight from end to end and exactly vertical, or "plumb." The masons

A finished concrete block wall.

stretch a taut piece of twine horizontally from one end of a row of bricks to the other to test for straightness before the mortar has hardened. They use a plumb bob to make sure the wall is vertical. A plumb bob is a weight, such as a piece of lead, attached to a string. The string is hung from the ceiling near the wall. When the weight is at rest (not swinging), the force of gravity makes the string exactly vertical, and the wall can be measured against it.

In service areas of the building, cinder-block walls would later be bare. But walls visible to the public would be plastered or covered with some decorative material.

As the concrete workers moved up to the higher floors, the ironworkers were reaching the end of their work. The building was about to attain its full stature. But for many of the contractors, work had just begun.

THE WALLS AND TOPPING OUT

By midsummer of 1970, the workers on the Exxon Building were in a state of total activity. Nearly everyone who would be working on the building was at work now. Those whose jobs were almost finished were near the top of the building, and those who were just starting were near the bottom. The different crews followed each other up the building in waves, finishing one floor and moving on to the next.

At the very top, more than 700 feet from the ground, the ironworkers were nearing the end of their job. Now they were working with steel members much lighter than those used on the foundation and lower floors. The girders were lifted from the ground by derricks and stacked on a temporary plank floor near the top. Then they were lifted once again and put into place. This was the same procedure that had been used below ground level. But now, high above the city streets, the dangers had increased.

A few floors below the ironworkers, the concrete workers were pouring another floor—the 48th. Below them, masons were at work and electricians, plumbers, and others were installing and connecting equipment.

On the 34th floor, another job was going on. The outside walls were taking shape. Above that level, the building was still just an open steel framework with floors that ended abruptly where the outside walls should have been. You could walk to the edge, lean over, and look straight down—if you had the courage.

The new concrete floor on the 34th story was already littered with the debris of construction. Plumbers and electricians were still at work. At the outside edges of the floor, near the drop-off, workers were busy installing the first part of the curtain wall—the thin slabs of limestone-laminated concrete that would form strips running from the ground floor to the top. Between these strips would be windows and glass-faced panels, which would be installed after the limestone slabs were firmly in place.

Each slab was a U-shaped piece of limestone backed with concrete. Although it would look solid when placed into position, the limestone itself was only two inches thick, and the concrete backing added another three inches. Both the slabs and the horizontal steel girders to which they would be attached were equipped with metal tabs that would allow them to be bolted securely together.

Several floors above where the slabs would be mounted, a derrick-like device had been mounted on a monorail. A cable was lowered from the derrick to the ground and a single slab was attached. The slab rose slowly on the lifting cable until it was at the right height. Then the whole lifting assembly—derrick, cable, and slab—moved sideways on the monorail until the slab hovered directly opposite its resting place.

Now on the 34th floor workmen watched another slab as it inched toward the building. Finally, they grabbed it and pulled it in until its tabs exactly fitted the tabs on the steel work. When it was in place, the workers bolted it into position, then freed the cable to descend and pick up another slab. Later, caulking men riding a scaffold outside the building would fill the seams between the slabs. When all the slabs were in posi-

The limestone-concrete slabs seem to wrap around columns or air-conditioning pipes. They are attached to the horizontal girders.

tion, each column would seem to rise 54 stories in an unbroken line.

Alternating with the limestone "columns" would be strips made of glass and special glass-faced panels that would hide the air-conditioning and other equipment placed in bays under each window. Although the panels are not transparent, they have a polished glass surface. From the street, they would look transparent enough so that there would seem to be solid glass strips stretching the height of the building between the strips of limestone.

Following the workers who installed the limestone by several floors, workmen arrived to install metal window frames between the limestone columns to prepare for the windows. Then the glass-faced panels were put into place. Now all that remained was to put in the windows.

The windows in the Exxon Building are each about 6 feet high and about $2\frac{1}{2}$ feet across. They are permanently fixed into position and cannot be opened. Because of this, the ventilating and air-conditioning systems are doubly important. Without windows to open for fresh air, the office workers will have to depend on the inside systems.

Putting these panes into the metal frames was less hazardous than placing the limestone, since the glass could be installed from the inside. There was no need to dangle it out over the workmen below. Still, installing the windows was a big job. There were more than 7,000 of them.

The walls of the Exxon Building were its most deceptive part. When they were finished, the huge columns of stone looked to some passers-by as if they held the huge structure up; yet they were only a few inches thick. The rest of the wall —the glass and the panels—was even thinner. Far from holding the building up, the outer walls were simply hung on the building.

Even when a floor was enclosed by thin, skin-like outer walls, it was far from finished. The floor was still bare concrete, there were few inside walls, and the ceiling was a mass of exposed pipes and wires. Masons had built equipment bays under the windows to cover the air-conditioning outlets, but they also looked half-finished, since they were made of rough concrete blocks or brick. The floor would not be ready for occupancy for many months.

The floor air-conditioning units and the windows are in place. Insulated pipes, partly hidden by a brick and concrete block wall, run up and down the building just behind the limestone slabs (see page 89).

Now the Exxon Building towers above its neighbors as the flag-draped top column is raised in the topping out ceremony.

At the same time, however, one important group of workers was finishing up. The ironworkers, who had put up the entire supporting skeleton for the building, were about to put up the highest piece of steel. Following a custom hundreds of years old, this event, called *topping out,* called for a celebration.

On August 7, 1970, representatives of the owners, architects, engineers, contractors, and unions were carried up to the 52nd floor. There they stood on the temporary wooden planking and watched as a two-story-long steel column with an American flag lashed to its tip was hoisted into place on the southwest corner of the building. The flag flapped lustily. Cameras snapped and officials grinned.

After the short ceremony, everyone left and went back to his job. After all, there was still much to be done.

LIFTING While the ironworkers were completing the skeleton and the lower parts of the building were taking on a finished look, another group of specialists was working out of sight deep inside the building.

They were the elevator installers. It was their job to put together the system by which people and supplies could reach the upper floors. It was a modern system of "vertical railways" which would be a miracle of efficiency and safety.

The building's planners had already decided how many passenger elevators were needed and how they should be arranged. With so many floors to be served, it was clear that no single passenger car should stop at every floor. Otherwise, a man who worked on the 50th floor might have to ride for ten minutes while the car stopped again and again on the way up.

The planners chose a system used in many skyscrapers. It called for four banks, or groups, of elevators, each of which would serve a nearly equal number of floors. The cars in the first bank would serve floors 3 through 14. Those in the second bank would pass the first 15 floors and would serve floors 16 through 27. The cars in the third bank would go straight to floors 28 through 39, and those in the fourth bank to floors 41 through 53. (Floors 15 and 40 are mechanical equipment floors not served by the passenger elevators.) This system allowed everyone to ride from the lobby to his or her own floor without changing cars.

The planners also decided that each bank should have eight cars, making a total of thirty-two passenger elevators for the building. This would assure that the cars would not be overcrowded and that passengers would not have to wait long for an elevator to their floor.

As the skyscraper went up, long shafts, or "hoistways," for the elevators were built in the center of the building. They were enclosed on three sides by concrete blocks. On the fourth side, an opening was left at each floor for the doorways. But until the elevators were built, the openings led only to a shaft which plunged all the way to the basement. From the top floors, one could see almost 750 feet down the empty shafts.

It was at the bottom of one of these shafts that the installers began their work. They were installing an elevator to serve the first fourteen floors. Naked light bulbs cast deep shadows on the concrete block walls as they put together the steel skeleton of the first car. Already attached to beams in the walls were rails which would guide the car up and down. The skeleton was nothing more than a hollow rectangle of steel. Small metal wheels were fitted on the outside of the car. These wheels would move along the rails on the wall, making sure that the car would ride up and down without joggling from side to side.

On top of the skeleton was a heavy metal rig to which six strong metal cables were attached. These cables would do the actual lifting.

When the skeleton was completed, it enclosed a box-shaped room. This was the car, or cab, in which passengers would ride. On one wall of the car, they installed a panel that would contain pushbuttons to control the elevator when it was completed. A bundle of electrical wires hung from the bottom of the car, and would connect the elevator's controls to the

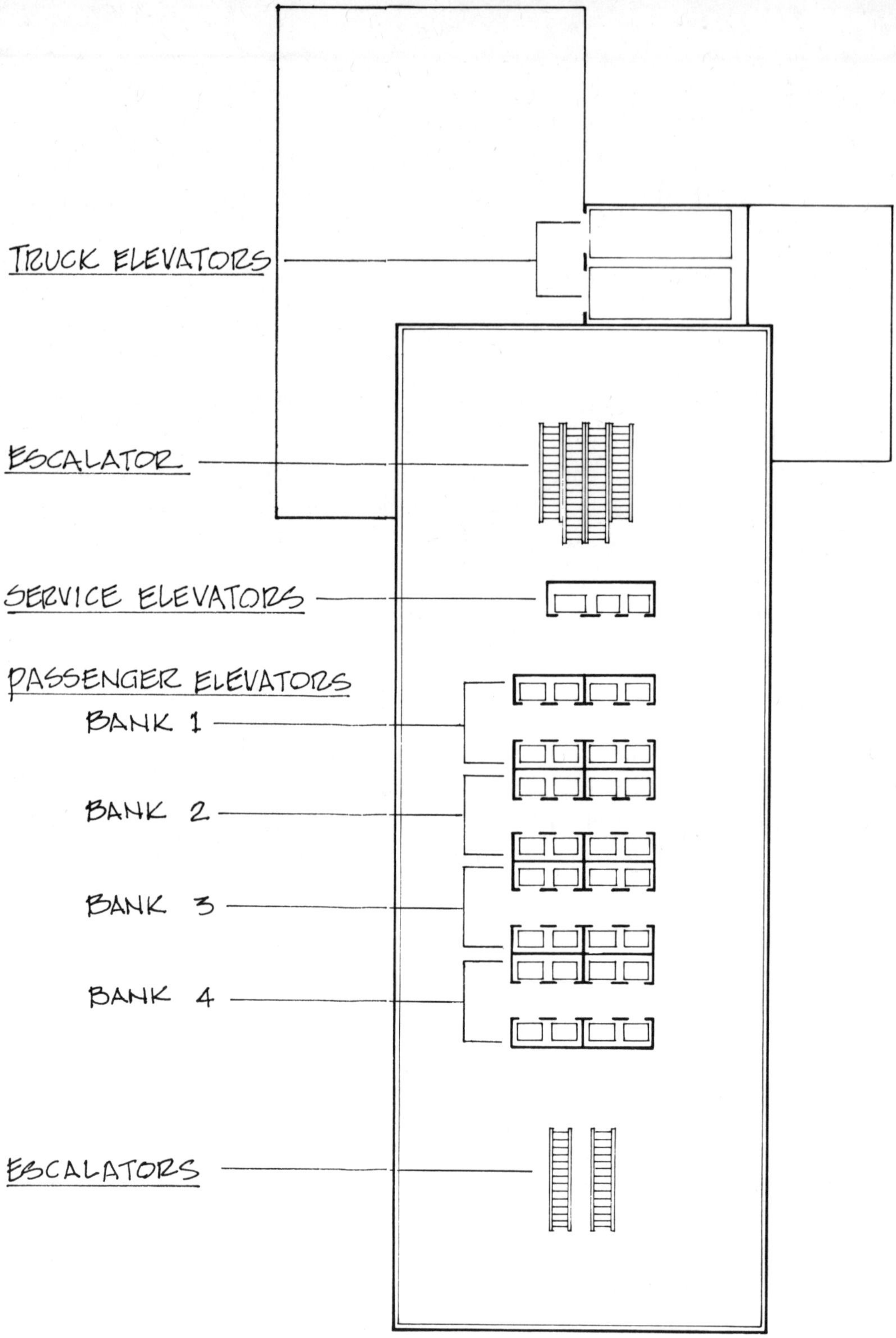

Above, a floor plan of the ground floor shows the lifting devices in the Exxon Building. Right, a cross-section of the building shows what floors they serve. Two truck elevators (T) lower trucks to underground loading platforms. Escalators (E) serve the subway concourse level and the mezzanine floor at the rear of the building. Three service elevators (S) serve all floors (the small doors in each elevator shaft show the floors on which elevators in that bank will stop). Passenger elevators are divided into four banks (1–4) of eight cars each. Bank 1 serves floors 3–14, Bank 2 floors 16–27, Bank 3 floors 28–39 and Bank 4 floors 41–53. The crossover floors (14, 27 and 40) allow a worker on any floor to go to any other floor without going all the way down to the lobby. The machine rooms (M) contain motors and electrical equipment for the elevators.

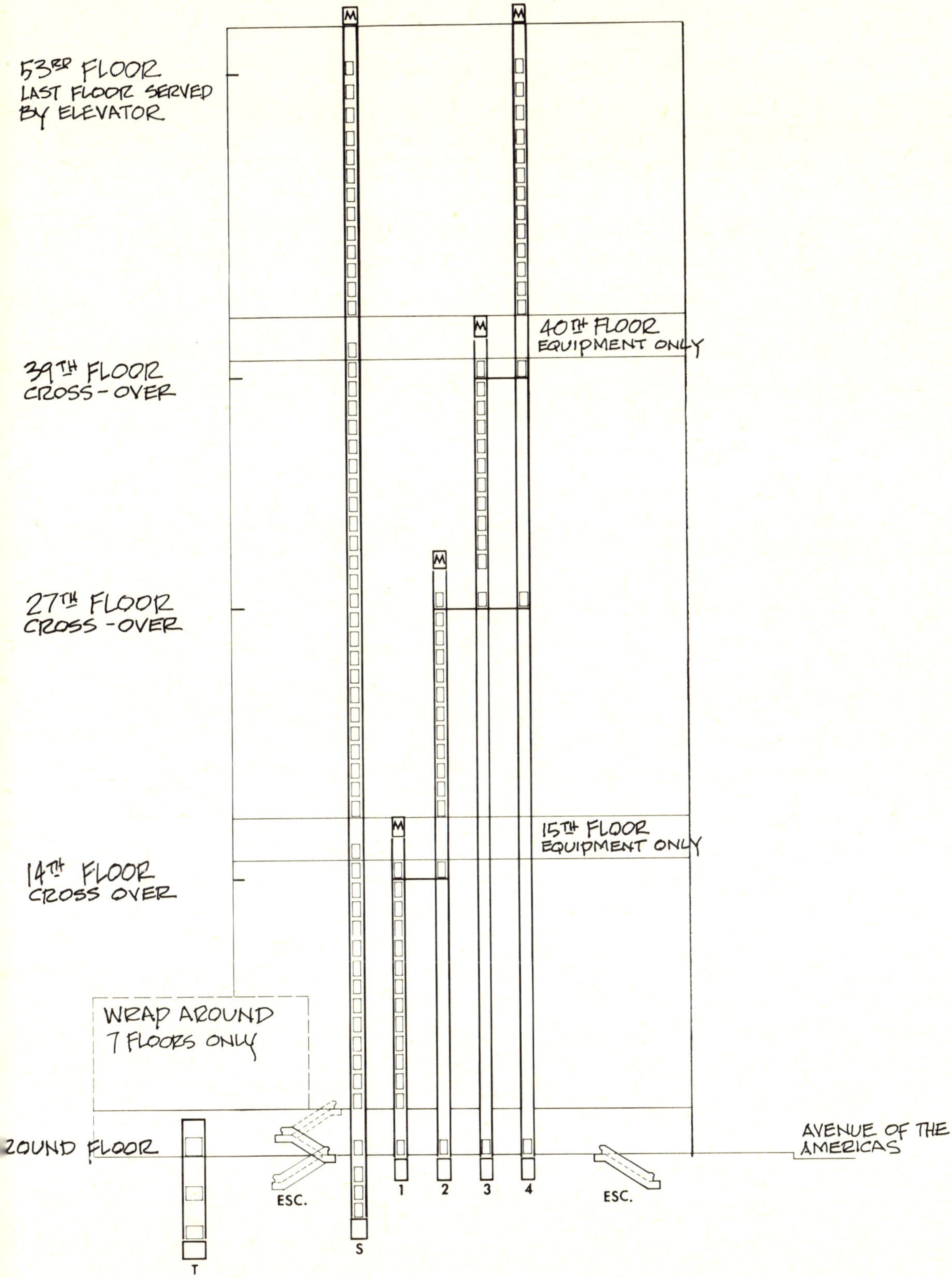
53RD FLOOR
LAST FLOOR SERVED
BY ELEVATOR
40TH FLOOR
EQUIPMENT ONLY
39TH FLOOR
CROSS-OVER
27TH FLOOR
CROSS-OVER
15TH FLOOR
EQUIPMENT ONLY
14TH FLOOR
CROSS OVER
WRAP AROUND
7 FLOORS ONLY
ROUND FLOOR
AVENUE OF THE
AMERICAS
ESC.
ESC.
T
S
1
2
3
4
M

motor at the top of the shaft. Finally, the men put in the doors that would open and close automatically when the elevator stopped on a floor.

Meanwhile, at the top of the shaft, on the 15th floor, another crew was installing the grooved steel wheel, called a *sheave*, over which the six lifting cables would run. A powerful electrical motor would turn the sheave and cause the elevator to ride up and down in the shaft.

The lifting cables are not designed to wind around the sheave. Instead, they just run over the top of it. On one side of the sheave, the cables are attached to the elevator. On the other side, they are attached to the counterweight, a massive piece of metal that weighs nearly as much as the car with a load of passengers. Special rails guide the counterweight along one wall of the shaft. As the elevator goes up, the counterweight goes down, and vice versa.

Since the car and the counterweight nearly balance each other, only a fairly small amount of energy is needed to move the car up and down. The better the weights are balanced, the less energy is needed to raise and lower the elevator.

The balance between the car and the counterweight was fairly equal, but another weight problem had to be considered. The lifting cables themselves were very heavy. When the car was near the bottom of the shaft, the weight of the cables was on its side. But when the car was near the top of the shaft, the heavy cables were on the counterweight's side of the sheave. Except when the car and the counterweight passed each other halfway up the shaft, one side would always be heavier than the other.

The elevator crew had an ingenious way to solve this problem. They attached one end of another set of cables to the bottom of the car and hooked the other end to the bottom of the counterweight. Now, as the car went up, it lifted more and more of these "equalizing cables," while the descending counterweight carried less and less.

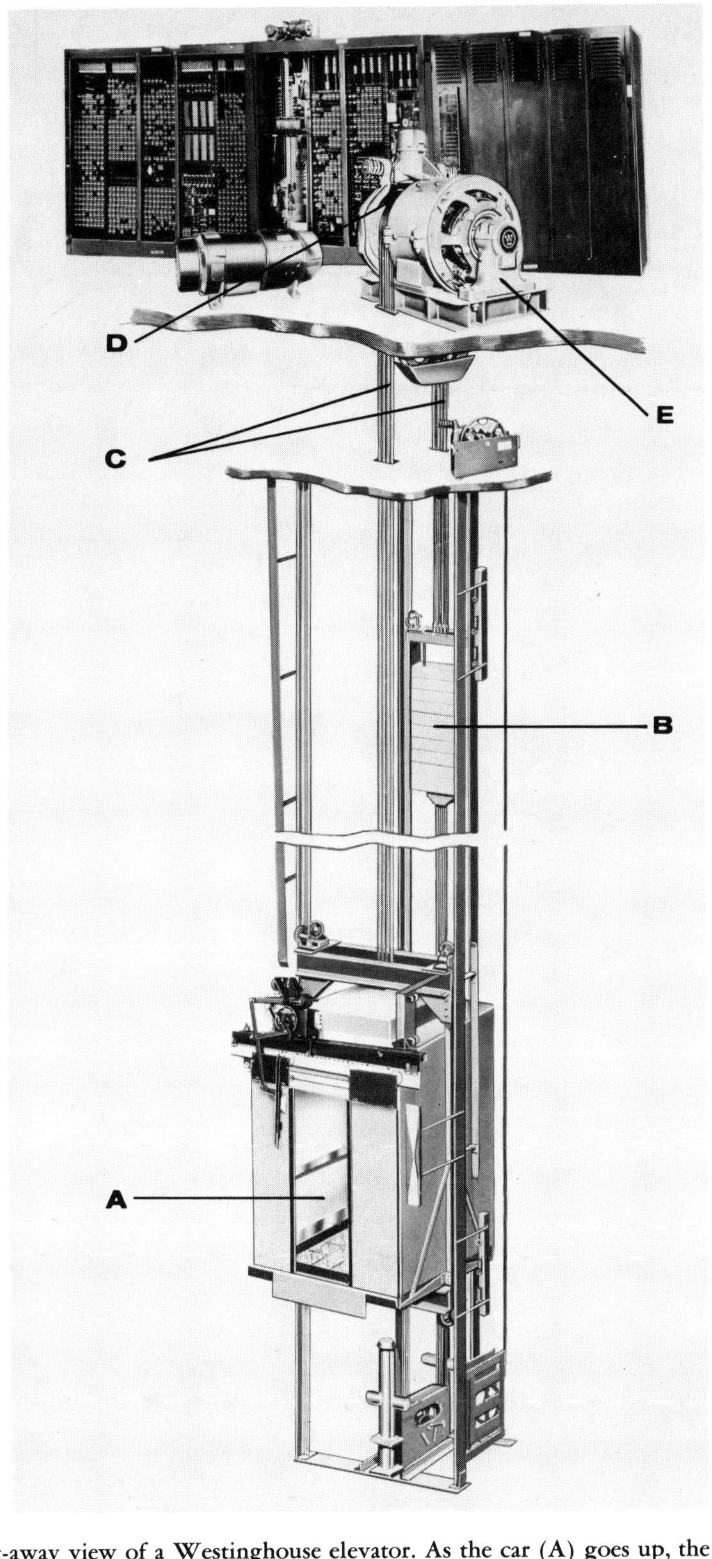

A cut-away view of a Westinghouse elevator. As the car (A) goes up, the counterweight (B) goes down. Both are attached to cables (C) which run over a sheave (D). The motor (E) turns the sheave and moves the cable to raise and lower the car. The electrical controls are in the background at top.

Until recent years, elevators were all run by a man in the car. Each operator received signals on his call board from passengers waiting to be picked up. But modern elevators need no operators. They are called "self-service" elevators, but that name tells only half the story. It is true that passengers push a button for the floor they want, but all the elevators are governed by a complicated control system.

The control system keeps track of all calls for elevators from every floor and adjusts automatically to answer each call and move the passenger to his destination as soon as possible. It "decides" who should be picked up first. For example, during the morning rush, when hundreds of workers are waiting

The sheave (left) and motor for elevator 23 in the Exxon Building.

for elevators going up from the ground floor, most of the cars pass by other workers who are waiting to come down. The control system "knows" that at this time of day more people will be waiting to go up than down.

A dispatcher in the lobby monitors the operation of the elevators on a huge control panel and is on hand if anything goes wrong. But in normal circumstances, the control system runs the elevators more efficiently than any single man or even a whole staff of elevator operators.

Part of the electrical controls for elevator 23.

The cars themselves are automated wonders, too. An express car can reach a top speed of 1,400 feet per minute. A passenger can travel from the ground to the 53rd floor in half a minute—fast enough to make his ears pop. Yet even at this speed, the cars are designed to start and stop so smoothly that the passenger hardly notices. (If a car rising at the speed of 1,400 feet per minute stopped abruptly, it would throw its passengers into the air.) The cars are programmed to stop precisely at each floor level. Their doors open and close automatically, yet they will remain open if anyone is between them. The elevators also have many other safety devices which make it almost impossible for passengers to be hurt in any circumstances.

The progress in high-speed elevator design has been a necessary step in making skyscrapers practical. Before electricity was harnessed, some elevators were run by steam engines. Others were actually raised and lowered by the men who rode them! In one of these slow, clumsy devices, it would have taken perhaps twelve or fifteen minutes to ride nonstop to the top of the Exxon Building. If it took such an elevator a half-hour to make a round trip, lines of waiting elevator passengers would probably stretch around the building. A worker on a high floor would have to spend hours waiting for and riding in elevators.

Lifting passengers is only half the job in a skyscraper. Office supplies are lifted by three service elevators that can stop at every floor. In addition, two special elevators were installed in the Exxon Building to lower complete trailer trucks from street level to a loading dock in the basement. These slow but strong lifting platforms are operated by hydraulic power. Powerful electrically-driven pumps drive fluid into a hydraulic cylinder, forcing a piston, on which the platform is mounted, to move upward. This same principle is used for the auto lifts often seen in garages and filling stations.

In the finished lobby, these elevators serve floors 3–14.

Other transportation systems in the Exxon Building include conveyors and pneumatic tubes. The conveyors move papers and supplies from one floor to another in special "tote boxes." The system is automatically controlled. An attendant dials the destination floor on the tote box. He then places the box in the conveyor system and it goes to its proper destination automatically. The pneumatic system carries messages in bullet-shaped carriers through tubes through which air is constantly flowing. The carriers race to their destinations automatically in a matter of seconds, making it possible to send written notes that are received almost immediately.

Most of the motors and control systems for the elevators and other transportation systems are housed on special equipment floors. Although they play an important role in the building's functioning, very few outsiders ever see them.

CONDITIONING THE AIR

Stepping onto the 40th floor—one of the Exxon Building's equipment floors—was, to a visitor, like stepping into a weird science fiction adventure. Although it was a brilliant September day in 1971, huge air ducts and house-sized boxes shut out most of the light. As the visitor's eyes got used to the dimness, he began to see smaller objects—pipes as big around as barrels and pumps the size of home furnaces.

Practically everything in sight was used to carry comfortable fresh air into the offices above and below, no matter what the weather was like outside. Two systems "conditioned" the air in the building. This system provided ventilation—fresh air. The other system helped to heat or cool the offices.

On the western end of the floor, one house-sized box after another stood silent and forbidding. These boxes, called *fan casings,* were covered with silvery metal foil. Huge sheet-metal air ducts that reminded the visitor of a roller coaster led to and from each box. The ducts overhead were big enough for a man to walk in. They ended in great curving funnels that led from the fan casings.

The house-sized boxes even had doorways. Stepping gingerly through one of these, the visitor found himself inside a rectangular room two stories high. One wall was pierced by several windows where air from outside would be sucked in.

The windows were covered with a sturdy wire mesh to keep birds and floating leaves and papers from being sucked in—and people from falling out. Inside the screen were big dampers, like oversized Venetian blinds. These dampers opened and closed automatically to let the proper amount of air pass through. This outside air was mixed with other air that had already circulated through the building.

Inside the fan casing it was unusually quiet. The walls were covered with a soft, dense matting to deaden the hiss and roar of rushing air. One side of the room looked like a mat of fine, dark green straw. This was a filter, and as the air passed

The 15th floor, a mechanical equipment floor, is still under construction. The huge air ducts and pipes at top are just being installed.

A door to one of the house-sized boxes is open, giving a glimpse of the coils through which air will be drawn in the conditioning process.

through, it would catch particles of fly ash and soot that pollute the city air.

The far wall in the second room looked like the radiator of a giant automobile. Like such a radiator, the wall was made up of a metal honeycomb. Hundreds of small fins were attached to the metal tubes that carried cooling or warming liquid. As the air was sucked through the grill, it would be heated or cooled to about 55 degrees, no matter how cold or warm it was outside.

Beyond the radiators, a smaller room contained a barricade of specially shaped nozzles which would inject steam into the air as it passed through, when it was necessary to increase the humidity. Just as the air had been warmed or cooled to a standard temperature, its relative humidity would be controlled. By the time the air reached the fan which had been sucking it through the rooms, it would be clean, cool (at 55 degrees), and just moist enough for comfort.

The fan itself was of the centrifugal type. It was a giant rotating drum that resembled the wheel in a squirrel cage. The blades of this spinning drum would draw the air in through the center and blast it outwards by centrifugal force into funnels that narrowed into the big sheet-metal air ducts. The pressure created by the fan would then force the conditioned air out into the main ductwork, pushing the air through a system of smaller and smaller ducts in the same way that water is pumped through pipes.

Finally the freshened air would emerge with a gentle hiss through ports concealed in the ceilings of hallways and individual offices. At the same time, another similar system would be sucking stale air out of the offices. This air would be returned to the 40th floor. Part would be blown out of the building and part would be "re-conditioned" and mixed with incoming fresh air.

This system is vital to the workers in the building. As we have learned, the windows cannot be opened, so there is no other source of fresh air. The conditioned air the workers receive is actually better than an open window would provide, especially on a hot, sticky day. But if the system should ever stop working, air circulation would stop and the air in the offices would soon become warm, stale, and unpleasant.

This air conditioning system, called the *interior* system, helps control the temperature in the building, cooling the air in summer and warming it in winter. But in order to have greater control of temperature, the building's planners called for a whole second system for the outside offices (those with windows). It is called the *perimeter* system.

In most outside offices in the Exxon Building, heating and cooling equipment was installed underneath the windows in bays made of concrete blocks or brick. A system of small pipes carries water—hot water in winter and cold water in summer—into a small radiator. Another system of air ducts carries conditioned air. When this air is blown into the room under high pressure, it causes the rest of the air in the office to circulate, passing the radiator, where it will be heated or cooled to the temperature selected by the tenant.

Although many outside offices have individual thermostats to control the temperature, the whole heating and ventilating system is controlled by a system that adjusts the supply of heat or cold to counteract the outside temperature, the effects of the wind, and even the changing position of the sun (which would bring added heat to offices on different sides of the building at different times of day).

The heating, cooling, and ventilation of the building require large amounts of machinery. The source of heat in the winter is steam, which is supplied by an outside company and is piped in. The steam transfers its heat to the water in the heating system. The warm water is then pumped through the building.

In the summer, water is cooled by huge banks of refrigeration equipment and then pumped through the building in much the same way. The refrigeration or cooling process is more complicated than the heating process, however. The circulating water collects heat from the air in the building. Then it returns to the cooling equipment. First the heat must be removed from the water. This is done by the same process used in refrigerators and home air conditioners.

The big problem is what to do with the unwanted heat. It can't be released at ground level or it would roast passers-by on what is already a warm day. The solution to this problem requires still more equipment and energy.

The heat is transferred to water in still another system of pipes. This water is then pumped nearly 800 feet straight up to the roof. There it is sprayed into the top of a huge cooling tower. As it drips down through the tower, huge fans 16 feet in diameter blow air past it. Part of the water evaporates and

A part of the building's cooling system, this refrigeration compressor in the sub-basement chills water that is then used to cool the air on hot summer days.

Looking directly down a pipe shaft in the Exxon Building. The toe of the photographer's shoe can be seen at bottom left.

blows away in clouds of water vapor. The remaining liquid is cooled by evaporation (just as swimmers are cooled when they get out of the water on a windy day). Then the water plunges back to the basement to pick up more unwanted heat.

On any hot summer day, steamy clouds of moist, warm air rise from the tops of modern skyscrapers. They are reminders that the people inside the buildings are probably cool and comfortable. The heat that was removed from the buildings is in the clouds themselves.

The various parts of the air-conditioning systems were installed at different times as the building went up. Much of the heavy heating and cooling equipment was below ground level. Other machinery was on three special equipment floors. In individual offices, the equipment—air ducts and water pipes—was installed in ceilings and in panels near the floor.

But by the time the office workers moved in, all of this equipment would be hidden from sight. Few of them would ever visit the workings in the sub-basements or on the equipment floors. And even in their own offices, the equipment would be almost invisible. Before they arrived, the offices would be "finished," transformed from roughness to elegance.

FINISHING

In late 1971 a number of new workmen were busy on the lower floors of the Exxon Building. When they arrived on a new floor, it still looked like a construction site. Exposed pipes and wires were visible everywhere. The concrete floor was littered with the waste and spills of earlier workmen. And the new workmen could see from one corner of the floor to another—it had not yet been divided into offices. When they left, the floor would be ready to use. These workers had many different specialties, but it was their responsibility to "finish" the building's insides.

When all the air-conditioning equipment had been installed under the windows and the masons had enclosed it with concrete blocks, a group of workers called lathers came to cover the brick with a heavy metal mesh called *lath.* On this lath the plasterers would apply a smooth, finished surface.

Plaster is a special combination of lime, sand, and water. Sometimes horsehair or other substances are added to give it additional strength. It has been used in one form or another for thousands of years. The plaster is mixed in big, flat tubs and is allowed to "slake," or set, overnight.

When the mix is ready, the plasterers apply it with the same tools and the same methods used by ancient plasterers. They spread it over the rough lath with a triangular or straight-sided trowel in a graceful, sweeping motion.

Time plays an important part in the plasterer's art. The rough coat must be allowed to set to hardness before the second coat can be applied. The second coat must also be allowed to harden, but it must not get too dry or it will absorb too much moisture from the final coat and cause it to crack. When everything is ready, the plasterer applies a final thin coat. How he smoothes it to a mirror-flat surface remains a mystery to all but other plasterers.

In earlier times, nearly all walls and ceilings in a new building would have been plastered. But in the Exxon Building and in many other new buildings, most walls and ceilings are "dry." They are made in factories and need only be installed. The walls are designed so that they can be moved easily, in case a company wants to change the floorplan. And the light, movable ceilings conceal pipes, air ducts, and electrical equipment that may someday need repairs. If the walls and ceilings were of plaster, rearrangement or repairs would be major jobs, requiring the walls or ceilings to be torn out and then replastered.

Before the finished ceiling was installed, electricians put in the large square fixtures that would contain U-shaped fluorescent bulbs to light each office. The ceilings themselves were false panels that hid the pipes and wires. First workmen hung a light metal frame, or gridwork, from the "real" ceiling. Then light fiber ceiling tiles were laid in the frames. In addition to hiding equipment and lowering the ceiling level, the tiles also helped to deaden sound, making the offices quieter.

The men who put in the walls between offices worked from plans provided by the new tenants, installing panels that were more than 8 feet high and as much as 15 feet long. These partitions were fastened into place by invisible snap-in devices at the floor and ceiling. Wherever a door was called for, they first put in a one-piece metal door frame. Later, other workers would come around to install the doors themselves.

Before any walls or other permanent fixtures are installed, careful measurements are made and marked on the unfinished concrete floor.

This left only the floors, which were still concrete. Now a flooring crew came through, spreading a tarlike substance called mastic and laying tiles of vinyl asbestos. When winter came, the workers would have to warm the tiles in a gas oven to make them lay properly. Then the tile floor was given its first heavy coat of wax to protect the surface from damage. In some corridors and offices, the floors were covered with carpet instead of tiles.

Finally the building was becoming an *office* building. Not only were the walls, ceilings, and floors in place; so were air-conditioning units, telephone and electrical outlets, lights, and everything else. The office worker would hardly be aware of all the planning, coordination, and skill that went into providing his own small work space. He would feel only the results—the convenient ride to his floor on the elevator, a comfortable temperature whatever the season, telephones, lights, even the quietness.

While the offices were being completed, other finishing work was going on elsewhere. Strangely, the ground floor of the building looked more unfinished than many of the offices inside. The ground-floor windows were left until last so that supplies could easily be moved into the building. The lobby had ceilings almost 30 feet high. Only now was it receiving its marble surfaces. The banks of elevators were still not complete, and work proceeded on these, too.

Then one day in the fall of 1971, great panes of glass began to arrive for installation on the ground floor. These windows would complete the building's outside appearance, and it seemed to the curious bystanders that the building would be finished when they were in place.

The installation was spectacular enough to attract a large crowd of "sidewalk superintendants." The sheets of glass arrived in a long truck, lying on their long edges. Each was 25 feet long and 8 feet wide. Obviously they were too heavy and too awkward to be moved by hand.

The truck was backed alongside the building next to where the first pane would be placed. Then a crane appeared inside the building and pushed its arm out through the opening. On the end of the arm was a long crossbar with suction cups on either end, each bigger around than a dinner plate. The crane was hand-operated—this was too delicate a job to trust to a machine.

First the big suction cups were applied to the top sheet of glass. To provide enough suction to lift the great pane, the air was sucked out of the cups by an electric pump, the only motorized machine used. Then a worker lifted the crane's arm and the pane, using a system of ropes and pulleys. A ground crew fixed hand suction cups to the pane so that they could help guide it. The crane was then pushed backward, moving the glass toward its destination. When the pane was near the wall, it was lowered carefully to the ground, still lying on its side, and leaned against the stone columns.

Workers now wheeled a pair of light metal scaffolds in front of the glass, one on each side of the window opening. Four workers scrambled up the scaffolding, two going to the top and two halfway up. They pulled clean rags from their pockets and wiped the metal slots where the glass would soon fit.

Now the crane operator pulled on his rope and the glass lifted slowly off the ground. As it rose, it was steadied by the ground crew with their hand suction cups. Then they began to rotate the pane into the vertical position. The men on the scaffold shouted warnings whenever the pane came near the

A corner of one of the lobby windows in final position.

building or the scaffold. It was too valuable a piece of glass to break.

As it turned slowly into position, the pane looked as if it were too wide for the frame, which was already in place. How would the workers ever fit it? But the planning had been done well—the fitting was a neat trick.

The glass was lifted so that it hung an inch below the top metal frame. The workers on the right held their edge of the pane while the workers on the left guided their edge carefully into the frame. This frame held the secret to getting the glass in. It was extra deep so that the pane could slide so far to the left that the right edge could clear its frame.

With one edge safely slotted, the workers on the right side, shouting instructions and warnings to each other, guided their edge neatly into position as the crane was moved ever so slightly to their side. Now both sides were safely in their slots.

It was a simple matter, now, to crank the glass up a few inches—with the workers at the top watching so that it didn't jam or chip—until its top edge slid into the top frame.

The bottom of the pane was still hanging in the air several inches above the metal ledge on which it would rest. Now the glass workers on the ground placed hard rubber wedges on the ledge. With a last round of shouting, the crane operator gently lowered the glass the last half-inch until it rested firmly on the wedges. The window was now in its final position. The suction cups were released, and the crane was pushed back out of the way.

All that remained was to fasten metal strips along the bottom of the window inside and out, and to caulk the seams, sealing the window into place. Then the procedure would be repeated for each of nearly 100 big windows that would ring the ground floor.

For the spectators on the sidewalk, the placing of the glass was a spectacular conclusion to their years of watching

After a huge glass pane has been installed, a workman on a movable scaffold caulks around the edges, sealing it into position.

A section of the finished lobby floor along 50th Street.

the building go up. Although it was not really finished, all the remaining work would be done inside, out of the sight of curious bystanders.

It had been a little more than three years since the excavation for the building was started. Now it would be only a few months before the building was ready for use.

MAKING THE BUILDING WORK

By the end of 1971, all of the basic construction on the Exxon Building had been done. There was no official completion date; as a floor became ready, the tenants for that floor could move in. Little by little, the building filled up with tenants. Now, in 1972, its life had begun.

In some ways, a modern skyscraper is like an ocean liner. The office workers are the passengers who use it, but don't understand its inner workings. Like a ship, a skyscraper needs a crew to keep it running smoothly.

As we learned in an earlier chapter, the Exxon Corporation is the major tenant in the building—it provides many of the "passengers." Rockefeller Center, Inc., manages the building. This company's "crew" operates and takes care of the building's equipment. The crew has three different responsibilities: *operations, maintenance,* and *security.*

The men responsible for operations must monitor all the building's complicated machinery and make sure that it is running smoothly. The man in charge of this effort is the chief engineer. His office in the sub-basement has a whole wall of dials and gauges that tell him at a glance how the building's major systems—ventilating, air conditioning, electricity, water supply—are working. In addition, he receives reports from the elevator "starters" in the lobby and from a team of inspectors

The chief engineer in his control room in the sub-basement.

who keep watch on the generators, turbines, pumps, and engines.

From his experience, the chief engineer can often tell when something is going wrong even before it has caused any inconvenience. If a motor begins to overheat, for instance, he may be able to send a crew to repair it before it grinds to a halt. When anything goes wrong—an elevator stuck between floors, a leak in a water pipe, or an overheated office—he hears about it and dispatches a crew to solve the problem. If the problem is a major one, he may have to call for outside help, but usually the building's own workmen can take care of it.

Other members of the crew are concerned with maintenance. They don't wait for something to go wrong; their job is to prevent trouble. On a regular schedule they inspect and service pumps, elevators, and other machinery. They look and listen for any potential problems—an overheated motor, a frayed elevator cable, a pump that vibrates too much or is running too noisily.

A member of the cleaning staff cleans an office after working hours.

The greatest enemy of machinery is friction. Wherever there are moving parts, they rub against each other, overheating and losing efficiency. Maintenance engineers lubricate such machines regularly. By applying a film of grease or oil, they reduce friction and keep the engines running smoothly.

Another big crew provides a different kind of maintenance. It is the cleaning staff which invades the building at night after the office workers have gone home. Teams of two cleaners work each corridor every night, emptying wastebaskets, cleaning desk tops, and keeping things straight. Other teams clean washrooms, clean and wax floors, and vacuum carpets on regular schedules. Without these workers, the building would soon become unfit to work in.

One special kind of cleaning is window washing. On older skyscrapers, where windows can be opened, window washers swing out hundreds of feet above the sidewalk, hook a leather belt to pegs on either side of the window, and lean against it while sponging off the collected dirt and grime. On newer buildings, the job is less hazardous. Two or more washers ride on a long railed platform which is hung from the roof by strong cables. They start at the top of the building, wash all windows within reach, and then lower the platform to the next floor. When they reach the bottom, they ride back up, the platform is moved over, and then they descend again, washing another strip of windows. On many buildings, they also wash the shiny panels between windows.

The planners of the Exxon Building, however, decided to try a new window-washing device—a machine that would wash the windows automatically. It rides in a platform which rides up and down the building on vertical tracks and is controlled by an operator on the roof. First the machine sprays the window with warm water. Then detergent is added to the spray, and scrubbing rollers loosen the dirt. After a rinse with clear water, the machine releases a blast of warm air to dry the window.

The automated platform is lowered by cables, and the machine washes two columns of windows at a time. One run up and down the building takes thirty-three minutes. After the platform is raised back to the top, it is moved over to the next columns of windows, and started again on its downward trip. There are 148 columns of windows in the Exxon tower, so the machine must make seventy-four trips up and down to complete the windows. Allowing for stormy days and repairs, the machine just about completes its job when it is time to start all over again.

A third part of the building's crew is the security force. Many of these men wear uniforms, and they patrol the building around the clock. It is their job to protect the workers and the property in the building. They watch for suspicious characters, direct visitors who are lost, and keep wanderers out of private or dangerous places. They also watch for any other danger to the building and its occupants—especially fire.

Looking down from the roof at the automatic window washer—and the street 750 feet below.

The Exxon Building was designed to be fireproof, and basically it is. Unlike a wooden building, it cannot "burn down." But fires can get started, and can still cause damage and injure people. Draperies, carpets, wastepaper, rubber cement, painting supplies, and other objects could provide plenty of fuel for a fire. The engineering staff is trained to help prevent fires and to act quickly if one should start.

One of the first problems in a building so large is to find out where the fire is. The security men who patrol the building might smell smoke and still spend precious minutes looking for the fire itself. For this reason, the building is equipped with an automatic smoke-detection system. In four locations on each floor there are automatic smoke detectors that constantly sample the air in different locations. If one of them detects an unusual amount of smoke, it lights an alarm on the panel in the chief engineer's office and sounds a buzzer to call his attention to it. The light tells him the exact location of the detector.

When this happens, he immediately calls the city fire department. Then he directs an assistant by walkie-talkie to the detector. The assistant investigates, and if there is any kind of fire he immediately pulls the nearest fire alarm (there are four on each floor). The alarm rings in a code that tells the exact location of the fire and sends the building's own fire-fighting team into action.

This team rolls a miniature fire engine out of its basement parking place into a freight elevator, which can usually get to the scene in minutes. The engine is equipped with chemicals that can even put out electrical fires, which cannot be extinguished with water. The city fire department is already on the way, too, in case it is a major blaze.

The greatest danger from a fire is not the flames or the heat, but the smoke. In case of fire, the ventilating system on the floor of the fire is shut off so that smoke will not be carried to other floors. Elevators are called to the lobby so that workers leaving the building will not be accidentally carried into the blaze. Fire wardens on each floor will help the office workers leave safely by way of the stairs.

In case of power failure, the Exxon Building also has emergency generators which will supply power to light stairways and lobbies and run pumps (for water) and one elevator.

The security systems are designed for events that would hardly be everyday occurrences. But the building and its crew stand ready just in case.

A SKYSCRAPER COMES ALIVE

It is a bright day early in the summer of 1972. A visitor finds his way through the plaza in front of the Exxon Building and enters the lobby through a glass revolving door. He is visiting a friend who works there.

The place is far different than it was a few years earlier. Instead of the clamor of heavy machinery and blasting, the visitor hears only the gentle rustle of office workers coming and going. Instead of dust and exhaust, he breathes air that has been purified, cooled, and dehumidified. Instead of descending into a giant excavation, he now rides an elevator to a floor high above the street.

Above and around him the building has become a reality. The architect's drawings, the engineer's figures, and the thousands of blueprints have now been translated into steel and concrete. After three years of work by thousands of men, the building is in use.

Although most of the work the building does is silent and invisible, a person with x-ray vision would see constant movement. Air is being sucked in, cooled, and circulated, and then blown out again. Elevators race up and down. Water is being pumped up to storage tanks and waste is plunging toward the basement, from which it will be sent out into the sewers. All over the building, pumps and motors whir, but the visitor senses only a calm stillness.

A view of the finished lobby's north side.

Inside the lobby, he studies the directory of tenants, realizing that there must be more than 4,000 workers already installed, with more to come. Right now, at the back of the building, a truck is arriving with desks, files, and other equipment as a new tenant moves in. Upstairs, a small city of executives, managers, receptionists, bookkeepers, secretaries, and office boys are tending to the affairs of dozens of businesses.

It is lunchtime, and dozens of workers are leaving the building. The neighborhood of the Exxon Building already had many large office towers, and the workers in the new building provide an additional load on an already crowded area. To help avoid overcrowding, the Exxon Building has its own concourse to the nearby subway station and restaurants to provide lunch for some workers right in the building. Many businesses have also agreed to stagger working hours so that some workers arrive early, before the rush hour, and leave early, while others arrive and leave late. This means less crowding in subways and streets.

The elevator doors open almost silently and the visitor steps aboard for his ride to an upper floor. Quiet music fills the car as it glides swiftly upward. When he leaves the car, the visitor finds himself in a luxurious lobby between two ranks of elevator doors. He has lost his sense of direction, but fortunately a receptionist is there to help him.

Soon a secretary arrives to show the visitor to his friend's office. As he follows her down the long corridors, he gets no impression of the underlying strength of the building and no sense of being hundreds of feet above the ground. The skeleton has been hidden behind attractive surfaces. Carpeting lies under his feet and richly paneled walls surround him. Even the activity inside the offices is hushed. Passing one door after another, he glimpses workers typing, filing, and talking on the phone or to each other. But the sounds they make are muffled by the sound-absorbing surfaces.

The visitor is shown into his friend's office, but the friend is on the phone. The visitor looks around. The office is almost like a den in a private home. Bookcases line the walls and drapes and Venetian blinds frame the large windows. The square light panels in the ceiling provide even light through the whole room. A faint hiss is the only reminder of the ventilating and air-conditioning systems at work to keep the office cool and comfortable.

The visitor feels a passing twinge of envy for the workers

An elegant office . . .

here. They are snug and secure, protected from wind, weather, noise, and even some kinds of air pollutants. They are surrounded by comforts and conveniences. Only the bleat of a distant taxi penetrates to remind the visitor of the world at street level.

. . . and a company cafeteria.

The visitor has lunch with his friend in a restaurant on a high floor of the building and then spends the rest of the day taking one last look around. He visits the roof and looks out on the other great skyscrapers among which the Exxon Building had gone up. Just south he notices two more buildings under construction, both resembling Exxon and both new additions to Rockefeller Center.

Later he stands again in the lobby as the hundreds of office workers get off the elevators and head for home—by subway, bus, car, taxi, or train. Even as they are leaving, the first members of the night maintenance crews are arriving. Although most of the building's "residents" go home after business hours, the building itself never sleeps. As darkness falls, the lights shine out on one floor after another as the cleaning crews dust, sweep, scrub, and polish.

By midnight the cleaning is done, but many of the lights remain on. All-night security men patrol the corridors, checking for prowlers, fires, or other troubles. The silence on the empty floors becomes deeper, punctuated only by an occasional telephone bell or the arrival of a service elevator.

At dawn, a new sound will arise—the air-conditioning units will begin preparing the building for the new day. At 7:30 the day crew of maintenance workers will arrive. Shortly afterward, the first office workers will appear, through the glass revolving doors or through the subway concourse. The elevator system will come alive, doors will open, and workers will speed up to their offices.

The great skyscraper will be ready for another day.

"As darkness falls, the lights shine out . . ."

INDEX

Page numbers in italics refer to photographs or diagrams.

PHOTOGRAPH CREDITS: Gil Amiaga for Rockefeller Center, Inc., front endpapers, 2, 6, 8 (bottom right), 8–9, 9 (bottom), 10, 13, 14, 29, 30, 33, 42, 44, 48, 50, 65, 83, 84; Bethlehem Steel Company, 47, 53, 59, 60, 68, 71, 73, 74; Louis Checkman for Rockefeller Center, Inc., 23; Tom Chin—Wagner International Photos, 9 (top), 77, 79, 81, 102, 105, 111, 117, 120, 123, 128; Edward Deeb—Kerby Saunders, Inc., 56, 63, 86, 87, 88, 91, 93, 107, 108, 112, 119; Impact Photos for Rockefeller Center, Inc., 1, 94; Norman McGrath for Welton Beckett Associates, 130, 131; Rockefeller Center, Inc., 8 (left), 20, 25, 122, 125, 133; Camilla H. Smith, 16, 55, back endpapers; The Smithsonian Institution, 37; Star Circle Industries, 115; Thomas Airviews for Rockefeller Center, Inc., 21; Gene Washnik for Westinghouse Electric Corporation, 101, 103.

Drawings on pages 27, 40, 41, 98, 99 by Thomas McHugh (based on original material from Harrison & Abramovitz & Harris Architects). Diagram on page 76 by Cal Sacks (based on original material from Rockefeller Center, Inc.).

Cover photos courtesy of Bethlehem Steel Company.